COLD BLOOD
HOT SEA

COLD BLOOD
HOT SEA
A Mara Tusconi Mystery

Charlene D'Avanzo

TORREY HOUSE PRESS

SALT LAKE CITY • TORREY

This is a work of fiction set in a real place. All characters in this novel are fictitious. Any resemblance to actual events or persons, living or dead, is entirely coincidental.

First Torrey House Press Edition, May 2016
Copyright © 2016 by Charlene D'Avanzo

Published by Torrey House Press
Salt Lake City, Utah
www.torreyhouse.com

International Standard Book Number: 978-1-937226-61-9
E-book ISBN: 978-1-937226-62-6
Library of Congress Control Number: 2015946056

Author photo by Derek Fowles Photography
Cover design by Rick Whipple, Sky Island Studio
Interior design by Russel Davis, Gray Dog Press

Distributed to the trade by Consortium Book Sales and Distribution

This book is dedicated to scientists struggling to understand extraordinarily complex phenomena associated with climate change. I was motivated to write *Cold Blood, Hot Sea* by stories of researchers maliciously targeted by climate change deniers.

I want to get out in the water. I want to see fish, real fish, not fish in a laboratory.
—Sylvia Earle, oceanographer and author

1

My father once said, "When you step aboard a ship, you leave solid behind for that vast unseen."

I rounded Maine Oceanographic's biology building at a trot—and stopped dead mid-stride. Goddamn, she was regal. Royal blue against the blood-orange April sky, hundred-fifty-foot research vessel *Intrepid* waited patiently for us, her mooring lines slack over yellow pilings.

Skirting a cart loaded with long skinny bottles, I all but skipped up the gangway and stepped aboard. *Intrepid* swayed with the incoming tide. My leg muscles tensed, and I checked the nausea patch behind my ear.

Yeah, I'm an oceanographer who gets seasick. Dreadfully, embarrassingly seasick.

With her new red offshore racing jacket and blond hair, Harvey Allison was easy to spot on the stern deck. She peered up at an orange buoy that looked like a gaudy mushroom lying on its side—a ten-foot-tall, thousand-pound one.

I backed down the ladder to join her. My slippery old sou'wester made shouldering my duffel awkward.

"*Dr.* Mara Tusconi," she teased. "Good morning."

"*Dr.* Harvina Allison."

Harvey reached up and ran her hand across bold black letters—MOI—Maine Oceanographic Institution. "It's almost as if the buoys can't wait to be released into the sea."

I patted the instrument. "Just like we've waited all winter for the temperature data these babies will collect."

Last year, ocean waters off Maine were the hottest in a hundred fifty years. Suddenly everyone who sold marine

1

critters—lobsters, shrimp, eels—demanded to know what the hell was going on.

"Now maybe we're more than nerdy scientists, Harv."

"The future of Maine fishing? That's high profile. Could be dangerous."

"What—?" *Intrepid*'s engines came to life and drowned me out.

Atop roiling water, the ship pushed seawater aside. Soon we'd depart. Legs wide for balance, I made my way to the port side and grabbed a handrail crusted with salt. I licked a bit from my palm and grinned.

Salt. My mother always said there was extra salt in my blood, because both my parents were ocean scientists. I looked toward MOI. If they were alive, Mom and Dad would be on the pier waving good-bye. They'd be proud of me. But they never knew I followed in their footsteps.

The ship surged forward and pulled away from Spruce Harbor's pier.

Harvey put her hand on my shoulder and interrupted my reverie. "Looks like a Winslow Homer painting from out here, don't you think? You know, wooden piers around the bay, lobster buoys, tree-covered hills."

The harbor blackened beneath a purple cloud. "Whoa," I said. "Mr. Homer's pissed off."

Harvey stared at the darkening sky. In profile her perfect features—high cheekbones, classic nose, large eyes—were even more evident. Who'd guess she drove a truck, had a rifle on her gun rack, and loved to shoot bear?

"Harvey, what did you mean danger—?"

The ship lurched. I heard a groan, and turned in time to catch a glimpse of bright orange shift behind a hydraulic crane. It looked as if the buoy might roll straight toward the starboard railing, an enormous toy top. Three guys jumped like fleas on a hot plate to stop it.

"Secure the lines!" someone yelled.

Crewmen scrambled to get the buoy back into position and secured it with stainless-steel cables.

"Bizarre," I said. "Gear that heavy not fastened tight?"

"Damn right. I'm first on the list to deploy," Harvey said. "Mine better not be that contrary buoy. Let's head down to our cabin."

Between us and the lower-deck staterooms were two sets of steep, narrow ladders. I faced each one and clambered down. The rhythmic drone of the ship's engine got louder and the stink of oil got stronger. By the time we reached our stateroom, diesel bouquet coated my tongue.

I threw my ratty duffel on the tiny desk next to Harvey's brand new one. *Intrepid* rolled to port and threw me onto the bunk bed. My stomach lurched, and I tasted diesel.

"Crap. I checked the weather forecast a dozen times. It's supposed to be calm until tomorrow. If the seas pick up, I'm in big trouble."

"Have your seasick patch on?"

I touched the spot behind my ear. "Yes. Look, I'm going up to check the forecast. And my email." I stepped out of the cabin, stuck my head back in. "Can't remember. Have you deployed buoys this large in rough weather?"

Harvey ran manicured fingers through her champagne bob, and looked at me, her gray eyes steady. "I'll be just fine."

I stepped out into the passageway and turned back once more. "You said dangerous. What did—?" But Harvey had already shut the toilet door.

A half-dozen computers ran along one side of the main deck laboratory. I slipped into one of the mismatched chairs. My dear friend Peter, the youngest PhD on board, clicked away at the keyboard next to me.

"Hey, Peter. How're Sarah and the twins?"

Focused on his computer, he furrowed his brow.

"Peter, what on earth's the matter?"

He held both sides of the monitor as if it might take off and turned toward me. "Bizarre email here. Hold on while I read it through."

I logged onto the NOAA weather site for the Gulf of Maine. A low-pressure system would bring squally weather faster than predicted. Winds fifteen to twenty knots, swells eight feet. My hand went to my stomach.

I skimmed my emails. The subject line "Climate Change Scientists Fudge Data" caught my eye. I leaned forward to read:

Email exchanges show climate change scientists create their own heat by cooking the data. The researchers' words—"transforming the data" and "removing outliers"—prove what the Prospect Institute has long known. So-called global warming is a manufactured fiction.

I turned to Peter. "Are you reading 'Scientists Fudge Data'?"

He swiveled his chair to face me. His dark eyes narrowed, black as the storm racing toward us. "Yeah. This one might get us."

"But everyone knows the Prospect Institute nuts claim smoking isn't a problem, there's no acid rain, ozone isn't depleted. They're not a credible source."

"They've hacked emails and quoted researchers' words. Don't you see? That's entirely different."

"Transforming data, removing outliers? That's just statistical lingo for data analysis. It doesn't mean we're fixing the numbers!"

I reread the message and stared at him, speechless. Like an athlete's doping scandal, this could ruin a scientist's career in a heartbeat. And the harassment could be horrific. In Australia, climate change scientists had to move after radical deniers threatened their families.

"There's something else, Mara. At the bottom of the email is a list of the ten hacked scientists. You're number seven."

2

My stomach lurched, but this time it wasn't the wild sea making me sick. "Why me? I'm not a famous climate change scientist."

Peter said, "Maybe it's your *Science Today* paper, and they've pegged you an up-and-coming troublemaker."

The bitter taste of bile filled my mouth as the room closed in. "I don't feel so great. Maybe we can talk later."

Back out on deck, my stomach settled down as I gulped cold sea air. I tried to quiet the chaos in my head. Could a bunch of quacks jeopardize my reputation? I wasn't an old silverback who could laugh off bad press. The funding I needed for research was hard enough to get. A scandal could be very bad news.

Peter might be right about my paper. I'd taken a chance with preliminary data and predicted unusually high temperatures in the Gulf of Maine this spring. As a young scientist, it's hard to get noticed. The irony was my desire for attention could have endangered my whole career.

I leaned back against the railing and looked around. A half dozen crew and scientists circled the buoys, peering at instruments. Harvey's deployment was soon, and I was on the list for the second one, after lunch. I tamed my long wind-whipped hair with an elastic band and twisted around. Twenty feet below, blue-green waves shot silver spray up the side of the ship.

Not good.

Someone bumped into me. A freckled redhead apologized, held out his hand, and pumped mine enthusiastically. "I'm Cyril, Dr. Tusconi. Cyril White. MOI photographer."

A redhead with a name like that. Cyril must've suffered as a kid.

"So happy to meet you, Dr. Tusconi."

I winced. Being called "Doctor" by a guy who looked sixteen made me feel my thirty-one years.

"Cyril, call me Mara."

"Cy. Rhymes with lie. Hey, I'll get better photos if I know what these buoys are for. I got the basics." He pointed toward them. "One end's the anchor, the orange float's on the other end, and instruments on top and below measure things like water velocity and temperature. But what's the purpose of this cruise?"

"To predict how ocean warming will impact Maine fisheries, we need measurements at more stations."

"Why?"

"Fishermen want to know if the ocean's warming. That impacts where and when they catch fish like cod. But Gulf of Maine temperatures vary a lot. The buoys give us better data, so we can judge if last year's highs were an anomaly or the beginning of a trend."

"Wow. This is a hot cruise. I'm psyched."

This *was* a hot cruise, and I was proud to be part of it. That the Prospect Institute might tarnish work critical to Maine fishing sent a spurt of outrage through me.

Cy was still speaking. "I went to your talk on climate change doubters. I had no idea. They harp on 'scientists aren't sure,' even though ninety-nine percent of experts agree—"

"Cy—"

This was the last thing I wanted to talk about at the moment, but the guy was on a roll.

"—the climate's changing and we're mainly the reason. They're going after scientists. Does that include you?"

I felt like he'd punched me in the stomach. "Where'd you hear that?"

He shrugged.

Time to change the subject. "If you were in my ocean-ography class, I'd give you an A. Shouldn't you be taking photos?"

He scampered off. For a photographer, he sure asked a lot of questions.

We'd reached the first deployment location. The engine droned down as the captain slowed *Intrepid* and held her steady on station. From the rear deck, Harvey shouted orders up to the winch operator. In her orange jumpsuit and yellow hardhat, she looked completely in charge. I gave her a little hand-pump.

The winch whirred, then kicked into a whine. Suddenly, a half-ton buoy sprang to life and lifted off the deck. Shipmates and scientists worked the guy wires to keep the buoy steady as it slid to the stern before dipping into its new home at sea.

Cyril White, back pressed against a portable van, tried to switch between the camera around his neck and cam-corder wedged between his feet. I walked over.

"Need help? I could hold the camcorder."

"Yes! Could you shoot the video?"

Intrepid rolled and nearly tossed me into Cyril. Given the sea state, videotaping wouldn't be smart. But this fraz-zled lad needed help. Cyril thrust the camcorder into my hands. I lifted the thing to eye level.

What the hell, I figured. It's just a few minutes. I cap-tured the orange blimp dangling from the crane, plunging into the sea, and popping up to whoops of the deckhands. In the water, the buoy looked like a half-submerged yellow R2-D2 topped with wind vanes and a couple of solar panel eyes.

Stupidly, I forgot a basic oceanographic physics lesson. *Intrepid*, now sideways to the waves, tossed port and star-board as well as fore and aft. I squinted through the viewer, trying to keep the bobbing buoy in the picture.

Bitter stuff oozed up from my stomach into my throat. I dropped the camcorder to my thighs, swallowed hard, took a deep breath, and lifted it once more.

The zooming back and forth with the camcorder did it.

I threw up. Bad enough. But I didn't do it over the side. I doubled over and let loose right where I stood. I'd had cereal for breakfast so, well, it was a goddamn mess.

Finally, my gastrointestinal track was empty. Panting and coughing, I blinked open tightly scrunched eyes. Splattered boots came into focus. I prayed it wasn't another scientist. Much better for a crewmember to see me in this thoroughly undignified condition.

I unfurled halfway. My splatter ran up yellow rain pants. *Please* be the deckhand who winked as I boarded the ship.

No such luck. I stood and looked into a chiseled face softened by wavy, straw-colored hair and lips turned up into a lopsided grin. Ted McKnight, my brand new colleague. Someone I'd really, really wanted to impress.

In a good way, that is.

He handed me a tissue.

I wiped my chin. "My god. I am *so* sorry."

His clear blue eyes flickered with amusement. "Hey, not your fault. Hold on a sec."

Ted skidded a water bucket my way, put his hands on my shoulders, spun me around, and splashed my rubber boots. I turned to face him, and he emptied the bucket on my boots and his pants. "There you go."

Before I could retort with something clever, Ted walked away to deal with the buoys.

One of the crew mumbled, "Great. A seasick oceanographer."

Ryan, first mate and my oceangoing pal, scowled at the seaman. "Enough." He turned toward me. "Don't you worry, Dr. Tusconi. We'll take care of this."

I climbed down the ladders to change. At the bottom, I missed a rung and landed with a thud at a crewman's feet. I looked up into the liver-colored eyes of a ponytailed bruiser who didn't bother to offer his hand.

"Ah, hi."

"J-Jake."

Jake walked away. If he were nicer, I'd feel sorry for a guy who stuttered his own name.

Sitting on my bunk, I stripped off the offending pants and pictured my excruciating moment with Ted. I pushed it out of my mind. No time for that now.

Back on deck, I closed my eyes and took in the cold, clean air. *Intrepid* was steaming straight now, heading for the next station.

The ship slowed to a crawl, and I popped my eyes open. The whistle sounded. Ear-splitting shrill—three short blasts—four times.

"Man overboard, port!"

I ran. The port railing was already three deep with scientists and crew. Crewmen below shouted as they lowered the rescue boat, but even on tiptoe I couldn't see them. From the railing, Ryan pointed past the aft end of the ship. "There! He's way back there!"

My mind raced through a grim list. If the man fell face down into the icy ocean, he'd reflexively gasp and flood his lungs with seawater. His blood pressure would spike. Then his heart would stop.

Frigid water was one reason why ship workers have the most dangerous job in the country. I put my hand on my chest and whispered, "God bless."

Ryan yelled, "Level the damn boat and get in!"

Finally, the outboard roared and faded into a drone as the men sped toward their target, probably hundreds of yards off by now.

The boatswain shouted, "Turn to!"

The railing cleared, and I peered over the side. The inflatable was heading back, a bright red dummy sprawled on her deck.

Ryan joined me. He yanked down his cap and shook his head. "Much too slow a drill. Guys looked like rookies."

3

THE SHIP RESUMED SPEED. IT was time for the senior scientists—me, Ted, Harvey, and Peter—to gather in the tiny lab off the fantail deck and review the deployment schedule. Regrettably, the meeting also included Seymour Hull.

Seymour, whom I'd nicknamed See Less Dull, was department chair and in charge of things that mattered, like grants. I refused to butter him up, and the man resented me.

He also could appear out of nowhere. "Mara, I need to speak with you."

I spun around.

Seymour's thin lips formed what could pass as a smile.

"Our meeting's now."

He held up my *Science Today* paper. "This will take a moment."

I waited.

He licked his lips. "Your paper."

"Yes?"

"You made a rash prediction and didn't pass it by me."

"Pass it by *you*?"

He waved the reprint. "Incorrect projections reflect badly on MOI. Not just you."

"Scientists sometimes make risky predictions. It's a judgment, and it's why they took the paper."

"I don't think so."

"What?"

"They published it because the author was a Tusconi."

I stepped closer and growled, "I do *not* use my father's name to get ahead. They took it because I'm an excellent scientist."

"Excellent?" He pointed to my nausea patch. "You can't even handle conditions out here."

I snatched the paper and marched toward the lab. That Seymour would throw my dead father's name in my face was obscene.

Seymour called out, "The Prospect Institute. More unwelcome publicity for MOI."

Harvey caught up with me. "That looked like a nasty interaction."

I quickly told her about the hacked emails and Seymour's accusation. "No suggestion I consult MOI's lawyers."

"He wants you to stew for a while."

"Yeah."

"And, Mara. What can you do about the email?"

"No idea. They don't teach you this stuff in grad school. I'll talk to Angelo when we get back."

Angelo de Luca, my godfather, is my only family. Twelve years ago my parents died in a research submarine accident. I was nineteen when my world fell apart. Angelo helped me try to make sense of the senseless and is as devoted to me now as I am to him.

He's my drift anchor in a rough sea.

We squeezed into the lab for our planning meeting. Head scientist, Harvey led the discussion. "We're on schedule with the deployments. Questions?"

"When can we look at CTD data?" I asked.

Tethered to the ship by high-strength line, the Conductivity-Temperature-Depth (CTD) profiler drops through the water and records real-time temperature and salinity from the surface down. Cutting-edge technology my parents' generation could only dream of.

"The profiler's already downloading," Harvey answered.

My throat tightened. What if—?

Peter asked, "Who's up for this afternoon's deployment?"

Ted gestured toward me. "It's Mara's turn."

"What's the report on that loose buoy?" Harvey asked.

We all turned toward Seymour, who shrugged. "Chief mate's looking into it."

Not a satisfying answer. Surprise telegraphed around the room.

"The buoy wasn't well secured," Ted said. "Are there inexperienced crew aboard?"

Standing near the exit, Seymour said, "I really don't know."

Peter met my look and raised an eyebrow. Figuring we were done, I was halfway out of my seat when he asked a question. I sat back down.

"I bumped into a guy in a passageway who's not at MOI. Who is he?"

Across from me, Ted fidgeted in his chair.

Seymour answered the question. "John Hamilton's a friend. He runs an aquaculture startup and is interested in our research. We had room, so I invited him."

A little odd but not worth ruffling Seymour's feathers.

"Seymour?" Peter said. But Seymour had left.

I looked at Peter. "Something wrong?"

Peter stared at the door. "Not sure."

Harvey and Peter filed out of the lab, and I started to follow. Ted said, "Mara, have a moment?"

The small space was littered with equipment, and we stood a few feet apart. Although MOI hired Ted two months earlier, he'd been in the Caribbean doing coral reef research. I'd barely spoken with him.

I'd forgotten how attractive Ted was. A bit of blond curled at his neck and his sunburned face looked in need of a shave. Both fair-haired and tall, he and Harvey would make a damn good-looking couple.

Ted had a good six inches on me, and I didn't want to talk to his throat. I stepped back a bit. "What's up?"

"The Prospect Institute email. Want to talk about it?"

I took a breath and felt my back muscles relax. "That'd be terrific. I've never dealt with anything like this."

"That message flabbergasted me. What're you thinking?"

"No time *to* think."

I waited for the seasick joke, but he didn't say a word.

"Maybe I should be proactive. You know, contact the papers."

"I have a close friend at the *Portland Ledger*," he said. "We were college roommates. If you want, I could talk to him. Or you could, of course."

"I'll think about it. Ted, thanks a lot."

He squinted and stared down at my eyes.

I tensed. Someone other than my ophthalmologist examining my eyes felt pretty weird. Maybe this was Ted's idea of a creative come-on.

"What?"

"Your eyes, Mara. They're really an unusual color. Forest green. What's the genetics—I mean, the color of your parents' eyes?"

Good. A scientific, not romantic, interest.

"Blue from my Irish mom and brown from Dad, the Italian side."

"A great combination." Ted's smile set off two faint dimples in his cheeks. His eyes searched mine in the normal way, and he left the lab.

I had to be careful not to assume the worst in men. The hurt of Davie's secret affairs was still raw after five years. Nevertheless, being overly suspicious wasn't a good thing.

Harvey stuck her head in. "You still here?"

"Ted asked about the email hacking. He also wanted to know about the genetics of my eye color."

"That's so t— um, so telling. I mean, that he's a scientist. You do have gorgeous eyes, Mara. With long auburn hair? Killer combination."

She scurried off.

I was sure Harvey was about to say "that's so Ted", which meant she knew him pretty well. Much as I appreciated her high regard, the comment about my physical features seemed like an intentional digression.

Back out on deck, I itched to check out the CTD data, but the weather was deteriorating fast. As *Intrepid* pitched from side to side, I made my way to the railing to inspect the sea state. Twice, I nearly tripped on a cable.

I grabbed for the railing and gasped. Row after row of steep-sided gray-green waves threw angry spray at the leaden sky.

Damn. Out in the brisk air, I felt okay. But straining to read numbers on a swaying monitor would not be wise.

Still, I *really* wanted to see temperature data. Just a peek. Legs wide for balance, I duck-walked to the lab, plopped in front of a monitor, and logged on.

I quickly found CTD data. My heart sped up as graphs appeared on the screen. Sliding side to side in the tethered chair, I leaned forward and squinted to make out the black temperature line. What was the scale? A ten—?

That was it. I'd crossed the Rubicon. In an instant, I felt dreadful—cold, clammy, faint. Stupid, stupid, stupid. I rushed out of the lab, sat down on a cable box, and drank in cold air.

Down in the mess, everyone else ate lunch and chatted away. Alone in a corner booth, I sipped hot tea and nibbled dry toast. First Mate Ryan walked up. Usually full of blarney, he touched my shoulder and spoke softly. "Dr. Tusconi. Can I get you anything? More tea?"

I looked up. In the shadow of his tweed cap, Ryan's blue eyes were all worry. "I'm okay, thanks. Hey, tomorrow I want to hear about your family's farm in Ireland."

Ryan wasn't gone a minute when Peter slid into the bench across from me. "You don't look so great."

"My patch isn't doing the trick today." I didn't want to

admit it was my fault for trying to read CTD graphs.

"Maybe it's something you ate," he said. "For this afternoon's deployment, I can take your place, no problem."

I blew out a breath and looked down at my hands. While accepting Peter's offer made sense, he didn't understand possible consequences. Of the twenty scientists and crew on board, Harvey and I were the only women. Others—Seymour, maybe even Ted—might think I was a girl who couldn't cut it. On the other hand, in my present state my reflexes would be slow. I'd be responsible for a half-ton instrument on a shifting deck with people around.

I reached for his hand and squeezed it. "Peter, that'd be terrific. Buy you a beer when we get back."

"You're on."

When we reached the next station, the weather was worse and sea rougher. Clearly, I'd made the right decision. I felt better in fresh air than anyplace else. So even though I couldn't supervise the deployment, I could join the others and watch it.

Out on deck, I found a good viewing spot under the winch platform. Above me, Ryan operated the winch that hoisted buoys up and dropped them into the water.

The angry sea made deployment tricky. Peter and the crew struggled to keep their feet under them, and communication was a challenge. Above the growing howl, Ryan and Peter shouted back and forth. Finally, Peter signaled he was ready. With a groan and a high whine, the winch came to life. Cable slowly peeled off the reel. As if waking from deep sleep, the buoy shuddered. Ryan played the winch so the hydro-wire tugged at the dead weight and inched the massive buoy up off the deck to ninety degrees.

The buoy was almost upright when Peter halted the operation. Reels droned down and stopped. Three guys in orange jumpsuits, legs wide, held guy wires tight and fought to keep the buoy steady as the ship did her best to topple them. Peter peered at the instruments one last time, then backed way. He signaled for the winch to slide the buoy to the open stern. Ryan powered up the reels and shifted gears.

Braced against the ladder, I was snug in my jumpsuit, wool hat, and fleece-lined boots. The wind's bite on my face, wet with sea spray and rain, faded as I fixed on the buoy's ride from ship to sea.

The taut hydro-wire held the buoy steady as Ryan slowly advanced it toward the stern. Peter yelled "Halt!" and Ryan slid the gear to neutral. Peter squinted at the buoy and frowned. Something about that buoy troubled him.

Peter signaled Ryan to power the winch once more.

What happened in seconds was slow motion to me. In frame one, the hydro-boom held the buoy upright, seaward of the stern. In the next, the ship pitched up. Like an enormous pendulum, the buoy swung back toward Peter.

4

PETER SAW IT COMING. He stretched out his arms as if he could deflect the tonnage, but his legs stayed glued to the deck. The bottom of the buoy hit Peter squarely in the chest. He fell backward and the entire thing rolled on top of him.

Peter's screams tore through the squall like a jagged knife through flesh. The captain sounded the alarm call. Frantic crew and scientists scrambled toward Peter, but there was nothing they could do. Ryan wrestled with the stuck winch, swearing a black streak like the Irish sailor he was.

It seemed like an hour but was probably less than a minute before the winch kicked in and lifted the buoy up and off Peter. He'd stopped screaming, and his leg below the hip was twisted at a sickening angle. Blood had undoubtedly pooled inside his jacket and pants, but only rain mixed with ocean spray ran across the ship's deck.

Two medics ran to Peter's side. Deckhands dropped to their knees and made a colorful semicircle around the stricken scientist. They waited while the captain and medical crew checked Peter's vitals. Peter looked peaceful, like he was asleep on the tossing deck.

Stunned, I held on to the ladder.

"He wants to talk to someone named Mara before we lift him up!" Coast Guard search and rescue hollered over the helicopter's roar sixty feet above. I scampered behind the man and knelt beside the rescue basket.

Shrouded and strapped, only Peter's head was exposed. In tight curls, his sandy hair was wet. The day after we got back, Peter was going to get a haircut, he'd said.

I bent over, my mouth next to his ear. "I'm here, Peter."

Peter's eyes fluttered open. I leaned closer. His words were slurred, halting, urgent.

"Not your fault, Mara. Not your fault."

Before I could respond, his eyes closed, and the strain around them faded as he slid into unconsciousness.

"Not your fault." What did he mean? Maybe that he volunteered to take my place. Or something was amiss with the buoy. I stood and stepped away from the basket. The medic gave the signal. As Peter rose off the deck toward the waiting helicopter, tears and sheeting rain ran together on my uplifted face.

The mess was nearly empty. Harvey, Ted, and I slid into a corner booth. A couple of crewmembers huddled together across the room stopped talking and between whispers glanced over at us.

Harvey sat next to me, head bent, elbows on the table. Again and again she ran her fingers through frowzy hair, trying to straighten out what couldn't be fixed. Unseeing, I stared ahead, replaying the movie of a buoy falling in slow motion.

Ted's voice broke in, gentle but firm. "We've got to talk about the rest of the cruise."

Seymour was on the bridge with the captain, so we could be candid.

Harvey straightened her back and coughed. "I keep thinking Peter will stroll in and joke about why he's late. He must be in the operating room fighting for his life."

I nodded. In the last hour, I'd roller coastered between shock, guilt, anger, and incredulity. I was spent.

Ted said, "We have to talk to the captain, of course, but should we scuttle the trip?"

Harvey leaned back and crossed her arms. "This is the second mishap in less than a day—by far worst. What's going on?"

We looked at one another. I said, "I have no idea. But wouldn't Peter want us to keep to the schedule and steam back tomorrow afternoon as planned?"

She put her hand on mine and squeezed it. "Yes. He would." Harvey kicked into gear. "Okay. We need a research plan which I'll pass by the captain. There are only two buoys left now. I'm scheduled to deploy rosette water samplers, but that's straightforward. Mara, do you still want to try out the new Video Plankton Recorder?"

"I'll wait on the VPR." The instrument was brand new, but even photos of microscopic critters taken right in the water didn't interest me now.

"I can supervise another deployment," Harvey said. "Want to do one, Ted?"

"If that's okay with Mara."

After what had just happened, I wasn't about to take any chances. I gave Ted a quick nod. "Thanks. And Harvey and I can handle the water samples." I touched the caddy at the end of the table. The salt and pepper shakers didn't rattle. "The storm's passing us now."

As Harvey and Ted discussed the winch malfunction, their voices faded into the drone of the ship's motor.

Harvey brought me back. "Mara, what are you thinking?"

"We've all heard about winch accidents of the past. Shutoffs didn't work, cables snapped, unsecured wire lashed across the deck. But that was before the regs."

Two lines formed between Harvey's eyebrows. "And?"

"Peter halted the deployment twice. He must've noticed something. And before the copter lifted him up, Peter mumbled 'not your fault.' So maybe he guessed it was some-one *else's* fault."

Harvey reached across the table and put her hand on Ted's. "Tell me again what Ryan said?"

Ted squeezed Harvey's hand and let it go. "The winch fouled. He freed it just as the ship pitched, and the buoy dropped. Which sounds like an accident to me. And the captain is calling it that, maybe a defective winch."

Harvey turned toward me. "Mara, do you have another idea?"

"What if there's inexperienced crew on board? Maybe the buoy-winch linkage wasn't set up right."

Harvey said, "You mean incompetence caused what happened?"

"Yes."

"The crew looked first rate for this morning's deployment," Ted said. "But can't we talk about all this back at MOI? We've got a lot of work to do, and we're down an experienced scientist."

Harvey got up. "I'll get up to the bridge and speak with the captain."

Ted slid out of the booth and stood next to her. It looked as though he was going to put his arm around Harvey's shoulder, like he was worried about her. "Want company?"

They walked out together.

I wished Ted had shown more interest in my guess about the crew and less in Harvey.

What followed was a long, long night. Thank god, the sea was calm.

There was much to accomplish and everyone—scientists, grad students, crew—rotated shifts to get it done. In and out of spotlights and shadows, we crowded the decks, and called out to each other. Over and over, we set and retrieved water sample arrays. We deployed the buoys with a new winch. In between it all, I hardly slept. I tried to catnap on my bunk, but all I could think about was Peter.

The captain kept us current on his condition—critical, no change.

At dawn, the sun erased a purple-red splash on the horizon above a placid sea, and by noon *Intrepid* was carrying its melancholy passengers home. I was about to tackle the first ladder down to the staterooms. A mousy man a couple of inches shorter than I am walked up.

"Dr. Tusconi, John Hamilton. Apologize for not introducing myself. With all that's happened—so terribly sorry."

Hamilton tossed his black knit cap from one hand to the other. With his deeply creased forehead, worried eyes, and downturned lips, the guy looked undeniably sad. I mumbled a few words of thanks. He nodded and walked away. It took me a minute to recall that he was Seymour's friend. He sure seemed a lot nicer than Seymour.

I was out on deck, duffle bag at my feet, when the first features of Spruce Harbor came into view. Actually, the Juniper Ledge bell buoy's clanging first announced the harbor's presence. In perfect order, houses and docks took shape, as if the town didn't yet know what had happened.

The ship passed the twin headlands protecting the harbor and left behind the buoys and what they'd tell us. Compared to the horror Peter and his family were going through, the April temperature data seemed a whole lot less important.

RV Intrepid slid alongside the institute's dock. As the crew secured the boat, I leaned over the rail and looked out at the venerable MOI brick buildings. My parents had worked in the same one that housed my lab and office now. The shadow of an idea drifted by and faded with a more immediate thought.

Harvey joined me. "You okay?"

"Just musing. Our building, my parents dead, maybe Peter dead. It looked like he was, you know, sleeping there on the deck."

A tall man with white curly hair jogged around the nearest building. He waved and called out, "Mara!"

Warmth filled me. "Angelo! I wasn't expecting you."

I turned to Harvey. "See you tomorrow. Get some sleep. You deserve it."

My godfather met me at the bottom of the gangway and led me to the side. His handsome face was pinched, and with dark smudges beneath them, his eyes looked cloud gray without the usual flecks of blue. Hands on my shoulders, he glanced down as if he'd worried I'd lost a body part. "I had to see that you were all right."

Angelo enveloped me, and I leaned into the softness of his down vest. My exhaustion and anguish gave way to tears. He let me go, and I stepped back and pulled myself together.

"My god, Mara. When I heard on the VHF about an accident on *Intrepid*. Well—"

"It was awful, dreadful. There's so much to talk about. But not now."

"I'll make a nice dinner tonight. We can talk then. Okay, sweetheart?"

That sounded perfect. Angelo strode away with a brisk step, and he waved at two seamen on their way to the ship. I watched until my godfather disappeared around the corner.

With the ship docked, we had to unload her. Even after a two-day trip, there was a lot to haul. Scientists and students marched up and down the gangway with crates of water samples, chemicals, computers—all the paraphernalia they'd brought aboard. Everything had to be moved to the loading dock, up the elevator, and into MOI labs. After it was stowed, people could go home to hot showers and meals with their housemates and spouses.

Except Peter, of course.

I did a final check of my lab. The microscopes were back in their usual places, computers reconnected, water samples stored safely in the freezer. All in good order, except for a flash drive I'd stashed in a drawer in *Intrepid*'s main lab. With a sigh, I schlepped back to the ship.

I stepped off the gangway. Two men caught my eye. Seymour was talking to a crewmate—liver-eyed Jake.

Seymour's face was inches away from Jake's nose as the crew-mate backed away. Hidden by a portable van, I slipped closer. The van blocked my view, but I could easily hear them.

Seymour growled, "Don't give me that duff. You're a clumsy fool!"

"But I mean that—"

"You mean? Mean what?"

Pause. Someone spoke in a calm, firm voice. Ted.

"Gentlemen, either of you need help?"

I backed off. The last thing I needed was for Seymour and Ted to see me spying on them. I made my way to the lab, pocketed the flash drive, and leaned against the counter. Seymour only spoke to the crew if he had to. But clearly Jake did something to make Seymour livid. Maybe the crewmate was somehow involved in the buoy disaster. If Seymour knew that, maybe he also knew why Peter kept checking that buoy—and other critical pieces of information.

Angelo De Luca is a widower who frustrates Spruce Harbor's older ladies. He's got a full head of hair—thick, white, and swept back—and a square chin sometimes darkened by stubble that gives him a rugged look. With the classic aquiline nose he calls beaked, his face would be at home on an old Roman coin. And, he can pull in a fighting bluefish no sweat.

But four years after they were married, Angelo's wife died in a car accident. She was twenty-five. Angelo says he'll never love another. Recently retired from MOI, Angelo is a brilliant marine engineer. In the 1960s, his oceanographic engineering teams designed instruments to help meteorologists make better weather models. That saved lives of Maine fishermen and boaters, some of them now his friends.

Angelo's home sits atop a bluff at the tip of Seal Point, one of Spruce Harbor's two headlands. At seven on the dot, my car splattered pebbles across the driveway as I swung to a stop. I reached for the door and hesitated, hand in mid-air. In order to make it on time, I'd driven too fast—and now wasn't ready to get out of the car.

Leaning back against the headrest, I deliberated in the shadow of the old gray shingled cottage that had been my refuge for the last eleven years.

I was uncertain about what to tell Angelo. I wanted him to know that Peter was injured by a buoy I was scheduled to deploy. But I had no evidence that inexperienced crewmembers might've been the cause. Maybe I was overreacting to what *Intrepid*'s captain guessed was an accident caused by a defective winch.

Start with the buoy, that's it. As a marine engineer, Angelo was the perfect person to ask why a winch designed not to fail did fail. Even better, Angelo helped lead buoy cruises on the newly acquired *Intrepid*. I stepped out of the car and walked briskly up the granite steps.

I kissed my godfather on the top of the head. He smelled of sea with a touch of olive oil. I took the opposite armchair in front of the crackling fire. He wore the wool vest I'd given him for Christmas. The blend of fibers picked up hints of blue in his gray eyes. A glass of wine sat on the wooden coffee table.

Like every good Italian, Angelo talked with his hands. Palms up, he gestured toward the solo glass. "Want some? It's Gavi, your favorite."

"Gavi would be perfect."

I leaned back in the chair and let out a long, slow breath. For the first time in days, I could let go and relax.

The living room—with wood planked floors and windows facing the harbor on one side and open Atlantic on the other—is my favorite. Growing up, I'd spent hours staring out those windows while Angelo and my parents talked about fish, fishermen, boats, and everything else to do with the sea.

Angelo returned with my wine and handed me the glass. "You seem far away."

"Oh, just picturing myself when I was little, nose pressed against the pane."

"The ocean was like a magnet for you."

I gestured toward the harbor. "Gorgeous sunset tonight."

Angelo nodded. For the evening's show, clouds on the horizon were slowly fading from vermillion to shifting mixes of purple. They'd soon turn gray.

We sat in comfortable silence—as people who'd suffered and taken care of each other can do. Finally, the only light came from dancing flames in the fireplace. Angelo got up and turned on a brass table lamp in the middle of an antique cherry table.

It was time to sift through the events on *Intrepid*.

Angelo slipped into his chair. "Any news about Peter?"

My throat tightened. "I called the hospital twice. He's unconscious in critical condition. I'll try again tomorrow." Staring at the dancing flames, I said, "I keep thinking about Sarah and the twins."

"Of course you do." He said quietly, "Tell me what happened."

I hesitated. Angelo's manner—careful speech and movement—signaled worry.

"I, well…"

"Mara, tell me. I need to know."

It tumbled out—the rolling buoy, the big sea. Peter taking my place, checking the buoy twice, crushed beneath it, airlifted to the hospital.

Angelo's face darkened with each revelation.

I finished, and he shook his head. "My god. What might've happened to you."

He looked away and pressed his eyes shut for a moment.

I brought him back to what he knew best. "What can you tell me about the winch?"

"What make is it?"

"Shapley Render-Recover's written on the side."

"It's new then, with top-of-the-line safety features."

I rubbed my arms, suddenly cold. "So an accident seems unlikely?"

Angelo hesitated, his brow heavily wrinkled. "I'd think so, but accidents *do* happen. MOI will work with the Coast Guard to find out. They'll want to interview you since you were right there."

I nodded. "Good. I want to tell them a few things that are, ah, peculiar."

"Yes?"

"I returned to the ship to retrieve a flash drive. Seymour was on deck dressing-down one of the crew."

"What did Seymour say?"

"Something about being clumsy."

"Anything else?"

"You know that little lab off the fantail deck?"

Angelo nodded.

"The senior scientists met there to review cruise details. We asked about the loose buoy, but Seymour didn't care."

Angelo frowned. "I've been on dozens of cruises and never saw a buoy roll on the deck. Seymour should've been alarmed."

"Right. At the time, the untethered buoy was bizarre. Now it's, I don't know, more than that." I hugged myself. "One more thing. Peter asked about a guy on the ship he'd never seen before."

"And?"

"Seymour said he was a friend interested in our work."

"It's unusual he'd be on the cruise, but it happens if there's space."

I slumped back in my chair.

Angelo stood. "Mara, you must be exhausted. Let's have dinner."

The Italian solution to any problem. Food.

Angelo announced we'd have shrimp simmered in his special tomato sauce poured over pasta. In the kitchen, I sat at one end of the long pine table and watched him cook. With the quick, practiced movements of a chef, Angelo sautéed garlic and onions in olive oil and added tomatoes.

Angelo's actions faded as images and emotions washed through me—crushing weight, smashed bones, Peter in the hospital, Sarah in tears, my own guilt and grief.

I sat back and closed my eyes. Angelo's right. I *was* exhausted. I drank in the aroma of garlic simmering in olive oil. After a glass of wine, the cruise was far away and unreal.

Angelo heaped two plates with linguini, shrimp, and sauce. I cleaned my plate and smiled at my godfather. "Guess I was hungry. That was terrific."

My cell phone rang. Harvey.

"May I come over?"

"Sure. I'm at Angelo's. What's happened?"

"Tell you when I get there."

Fifteen minutes later, Harvey sat at the kitchen table. Her eyelashes were wet, face puffy. I held her hand and braced for the worst.

It was barely more than a whisper. "Peter died an hour ago."

5

"PETER'S DEAD?" OPEN MOUTHED, I stared at Harvey.
She dabbed her tears with a damp linen handkerchief.

Angelo murmured, "My god."

"Massive internal injuries. Nothing the doctors could do."

"How did you find out? Isn't he down in a Portland hospital?"

"My niece is an intensive care nurse. Sarah was down there, of course, and said she wanted us to know."

"Give me a minute."

I stumbled into the living room, over to a window. Dizzy, I leaned against the pane and closed my eyes. When I opened them, the channel buoy's warning light flashed red on the black sea.

Angelo slipped next to me. "I'm so very sorry about Peter."

I held onto the sill and looked into the night. "A husband, dad, friend, young scientist. Gone in an instant." I turned toward him. "It feels all too familiar. Mom and Dad, you know—"

Angelo opened his arms and pulled me close. I rested my cheek against the soft wool of his vest and drank in warmth and succor. I stepped back, blinked, and whispered, "I love you, Angelo."

"I'm always here for you, Mara, you know that. Peter and your parents—two oceanographic accidents? It's not surprising his death triggers old memories."

I coughed. "You're right. We'd better get back to Harvey."

In the kitchen Harvey stared into space. As usual, everything about her—clothes, hair, posture—was perfect. But something about her gave the impression that, like a glass statue, she could fall over and crack with a little push.

I slid into the chair next to her. "Can I get you anything? Coffee, tea?"

She murmured, "Just a glass of water."

Angelo let the tap run and placed glasses of cold well water in front of Harvey and me. Between sips, we talked about Peter and his family. Harvey was a good friend of Sarah's.

She leaned against the table and stood. "I'm exhausted. I'll call Sarah tomorrow."

I walked Harvey to the kitchen door. We hugged and held onto each other's hands.

"Harv, I'll call you first thing in the morning."

She dropped my hand, turned, and walked down the stone steps. I expected her to do the usual reverse hand wave over her shoulder. But she didn't.

In the kitchen, Angelo lit a burner on his ancient gas stove and got out the Italian roast. The teakettle whistled. He poured steaming water into a press and coffee aroma filled the kitchen.

Angelo slid a mug and bakery box in front of me. "Decaf and biscotti. You must be all in."

The sugar perked me up a bit, but sorrow and guilt had taken their toll. "I'll be off in a few minutes." I nibbled the biscuit. "When I'm interviewed, what will they ask?"

"Why Peter took your place, what you saw. That type of thing."

I dunked the rest of the biscotti. "Ted was there too."

"Ted?"

"New hire. Algae expert. Ted McKnight." The last thing Ted and I talked about was the Prospect Institute email. I opened my mouth, then shut it.

Angelo tilted his head. "What?"

He didn't need more to worry about. "Nothing, really."

"You're worn out, dear. Why don't you head home? Get some sleep." He followed me out. At the bottom of the steps he said, "Mara, please stop by tomorrow. I'll worry about you."

On tiptoe, I kissed him on the cheek and promised I would.

Angelo was right. Fifteen minutes after I got home, I fell into bed and slept seven hours straight.

The next morning, I waded through the cruise data at my office desk. I'd called Harvey who said she'd get back to me about Sarah—so making graphs was a welcome distraction. A few hours later, I finished a preliminary look at the temperature data. Someone knocked on the door.

"It's open!"

My cousin, Gordy Maloy, stuck his head in. His ratty Maine Fishermen's League cap shadowed intense blue eyes that didn't miss a thing. "Doc, got a minute?"

As with any native Mainer, this came out, "Gaut a minut?"

"Always for you, Gordy."

Even though our mothers were sisters, Gordy's upbringing was very different from mine. Bookish Bridget, my mom, became a marine scientist and married Carlos Tusconi, an oceanographer. Kate, Gordy's mother, fell for an Irish lobsterman from Lubec. Gordy grew up in northern Maine around folks who made their living off the land and sea, while I spent my childhood around intellectuals like my parents.

He pulled off his cap and tossed it on my desk. "Heard 'bout the cruise. Damn shame."

"What's the buzz?"

"One o' yours died. Buoy accident."

I pointed to a chair. "Not why it happened?"

"Ryan O'Shea was on the winch. Some's sayin' he's accountable. That's a load of bollocks."

I nodded. "Ryan's a fine man."

31

"Got that right." He looked down at his feet. "Um, boys in the League's been pesterin' me to ask 'bout the water temperature, but—"

"I'm working through it now. Here's some data from a station off Mount Desert, on the continental shelf." I swung the monitor in his direction and pointed to a graph.

He leaned forward for a better look. "I fished 'round there. So what's this?"

"Last year. Temperature down through the water."

He ran a finger across graying square-cut sideburns. "Gimme a sec. It's from the surface down fifty feet, and, um, somethin' like five degrees hotter at the surface."

"That's it," I said. "Last April, the surface water was six degrees Celsius, about forty-two Fahrenheit. The highest ever measured."

He whistled. "I didn't know it was *that* high."

"For fishermen, what difference would that make?"

Gordy stepped back. "Temperature's everything. Take cod. If water's too warm, their eggs'll die. And where they go, so where we fish, depends on the temperature. Good fishin's moving north, we know that. They're off Georges Bank, we're screwed."

I nodded. For over four hundred years, New England fishermen had pulled cod and halibut from George's Bank's fertile waters. Losing that fishery was unthinkable.

"But that's las' year," Gordy said. "What 'bout this year? That's our bet."

I inserted the graph I'd made.

Gordy leaned in and used his finger to draw the temperature profile in the air. It took a minute for his pursed lips to spread into a wide grin. "See? It's back ta normal, jus' like I said. Cousin, looks like you owe me a bottle of the best scotch they sell in Spruce Harbor."

"You'll get the scotch. But it's still early April. In a week or so, I expect the surface temperature will go right up. Like last year, just later."

He reached for his cap and yanked it on. "Ya know, warm water's bad. But it's almost like you want it to happen."

That was Gordy. Called it like he saw it. "It's a problem with ecology research. As if we're looking for disasters. I don't *want* Maine waters to get warmer. But since everything points in that direction, we need the best data we can get to show what's going on."

After Gordy left, I ran a finger over the keyboard. There was a lot riding on my assumption that springtime waters off Maine would be as warm, or even warmer, than the previous year. So far, the data didn't show that. It *was* early days, I told myself. The buoys would continue churning out numbers for months.

But Gordy was right. A repeat of last year's high temperature could be bad for fish and fishermen. And here I was looking for it to happen. It was good he was around to remind me about that.

I sighed and arched my back. A critical grant proposal date loomed, and mine centered on the idea that Maine's spring ocean temperatures were regularly high. For researchers like me, getting grants was imperative. To study Gulf of Maine warming, I needed money to go to sea, buy equipment, and pay graduate students. That money came from sources like NOAA—the National Oceanographic and Atmospheric Administration. The NOAA proposal was due in a few weeks.

I needed to begin writing it now, but if the buoy temperatures didn't rise quickly I'd be in trouble.

Two days ago, this realization would have devastated me. Now it was something I simply had to face. The same went for my *Science Today* paper. My prediction about even warmer water this spring might be wrong. I had to put that worry aside and figure out how to deal with it later.

Funny how death puts things into perspective.

I took the stairs down to the Biological Oceanography office and learned that Peter's memorial service would probably be early the following week. Seymour wasn't there. I grabbed a cup of coffee and left before he appeared.

Back on the third floor, I was partly down the hallway when a plug of a woman in a plaid flannel shirt and army boots reached the top of the stairs. Betty Buttz—cantankerous, smart as a whip, emerita professor at MOI.

I called out, "Betty. Got a minute?"

Betty kept marching and threw a response over her shoulder. "What?"

I didn't mind. Betty was equally cranky with everyone.

I caught up with her. "You heard what happened on *Intrepid*?"

"'Course. It's all around town."

"Can we talk?"

Betty scrunched her mouth, as if about to say no. Instead, she invited me into her tiny office. The walls were lined top to bottom with thick books, some with faded titles. She motioned to a folding chair.

"What is it?"

I told her everything that had happened. "You've been around winches for a long time. What do you think?"

Betty shook her head. "The winch freefalling like that. Suspicious, but hard to know."

"And MOI's investigation? Any thoughts there?"

She sat back and crossed her arms. "Think about it, honey. They're gonna shift the blame to the winch manufacturer as fast as they can. Bad publicity, especially *that* kind, is a disaster for this institution. Oceanographic research is extremely expensive, donors skittish."

"MOI? You can't mean they'd cover up wrongdoing."

"No, but they may not dig too far, if you get my meaning."

I let Betty's words sink in. "How can we find out what happened?"

"Damn arthritis." She shifted her position and stretched her right leg. "If it was me, I'd investigate on my own. But not so anyone knew."

"Boy, I'll have to think about that. I just want to do research, teach, work with grad students, and write."

She shrugged. I waited for more, but Betty was a woman of few words. Unless, I discovered, they were about Seymour.

"As long as I've got you, I'd like your take on Seymour."

She rolled her eyes. I took that as a signal to keep going. "For one thing, I'm beginning to wonder about his competence. He should've been more concerned about that loose buoy."

She snorted. "Seymour's had an iffy reputation for a long, long time."

I leaned toward her. "Fill me in."

"Twenty years ago, Seymour was a cancer researcher down in Woods Hole studying squid. He worked with someone. Can't remember who. Anyway, word got out that he wasn't an independent scientist. Phillip Morris paid for his research. Seymour was an expert scientist for tobacco interests."

I whistled. To me, this was like being paid for sex. "Was there evidence he cooked his data?"

"No proof. In those days, electronic data trails were hard to trace. But I think he did. His cancer tobacco research was just too clean. Anyway, that's why I put up a stink when he applied for the MOI job."

"So why was he hired?"

"He's aggressive as hell and knows how to get dough. Like I said, oceanographic research is expensive."

I nodded. Seymour brought in a lot of money for MOI.

While I was at it, I figured Betty could help me understand what really bugged me about him. "As you know, Seymour was here when I was hired."

"I remember. You came right from your post-doc and looked like you were fifteen. This was your first real job."

"Right. And department chairs support and mentor young faculty. But from day one, Seymour's been snarky to me. Any idea why?"

She rotated her right ankle. "I bet it's something to do with your parents. As department head, he's automatically Distinguished Professor of Marine Science. He's probably sick and tired of hearing about the brilliant Tusconis."

That made sense. After my parents' deaths, MOI established the distinguished professorship in their name. To be fair, repeated remarks about the wonders of two people you never met could get old.

I dearly loved my parents, but sometimes their legacy was a burden, even for me.

"Thanks for everything. Let's share a pot of tea at the Neap Tide soon. Bet you know a few stories about my parents."

She looked up at me. "Sure. You seem like a smart cookie, Mara. But watch yourself with Seymour. If he thinks you suspect him of anything, he'll come after you like a two ton truck."

By the time I reached my office, I felt uneasy. Being the target of Seymour's spite was a dreadful prospect. I shut my office door and plopped into a chair. The phone rang.

Sharon from the Biology Department office said, "Chief Warrant Officer Wilson called. He's Coast Guard and wants to ask you about the *Intrepid* incident. Do you have time? He'll come to your office."

"Sure. Now's okay."

Fifteen minutes later, Officer Wilson, spotless in his whites, was seated in my office. My spare wooden chair was too small for his two hundred pound bulk, and each time he shifted position his Coast Guard cap fell off his knee. With a quick look at the floor, he shoved the thing under his armpit.

I answered his questions as clearly as possible. Peter took my place because I was seasick. Yes, I was out on the deck when the buoy dropped. No, I didn't know why Peter halted the deployment or what he was looking at on the buoy.

Wilson stood and straightened up so tall I thought he was going to salute me.

"That's all, Dr. Tusconi. I appreciate your help with this accident."

I stood as well. "Accident? You think it was an *accident*?"

He smoothed his hair and settled the cap into proper position. "Sorry. Can't say more. It's an ongoing investigation. We're talking to everyone."

The man was out the door before I could respond. The whole interview couldn't have taken more than ten minutes.

Harvey appeared in my doorway. She looked toward the stairwell. "Who was that?"

"Lots to tell you."

She stepped into my office, and I closed the door. Harvey had come from a run and looked as sleek as a cat in black compression leggings and a hot pink wind jacket. Her high-end earphones only added to the allure.

"Cool duds," I said. Her outfit was better suited for Central Park, not Down East Maine, but I admired Harvey enormously for her class and style.

She grinned. "Too bad you're my only admirer." The smile faded. "Did you sleep? How're you doing?"

"I'm okay. 'Course, I keep thinking about Peter's family."

"Me too. Running helped a bit."

We were silent and sad for a moment.

"I just got an earful from Betty Buttz about MOI and Seymour. And right after that a Coast Guard officer interviewed me."

"Think I need to sit down for this."

"Betty thinks MOI might not look too hard into Peter's death. They'll try to shift the blame somewhere else, like the winch company."

"Because of bad P.R.?"

"And fundraising."

"Betty can be awfully cynical. But she might be right."

"When we spoke I wasn't sure. But now, I'm beginning to think she is. This Officer Wilson called what happened an 'accident.'" I air-quoted the word. "I questioned that, and he gave me the 'we're talking to everyone' line. Maybe I'm overreacting, but he couldn't get out of here fast enough. The Coast Guard's in charge of things like this. You'd think they'd do a better job."

"But what can you do about it?"

"Betty thinks if I really want to know what happened, I might have to investigate myself."

She sat up even straighter than usual. "Mara, you're no detective."

"Scientists work on puzzles all the time. I can apply scientific reasoning to this one."

Brows knit, she studied my face. "What's at the bottom of this? Guilt?"

I walked to my floor to ceiling window. An osprey skimmed over the water, searching for prey.

I turned toward her. "Sure I feel guilty—the whole thing's horrendous. But it's more than that. Peter was a good friend. The best. His death deserves honest and thorough investigation."

"But what does that *mean*?"

"It means an authentic, in-depth look at the circumstances. Maybe start with weird things that happened, like the loose buoy"

"Do you have time for this? What about your work?"

"I'll have to fit it in."

"And not sleep." Harvey shifted in her chair. "Seymour and *Intrepid*'s captain. If they claim it was an accident, you'd be challenging them."

"Yeah, but I just feel. I don't know. Damn it, I've got to do *something*."

She shook her head. "This feels too risky. Seymour's already on your case."

"Yeah."

"And if you pursue this, I can't help."

"I didn't ask—"

"We've talked about me being department chair after Seymour."

"And this might jeopardize your chance."

She stood. "Hope you understand."

Harvey left, and I frowned at the closed door. I couldn't remember the last time Harvey and I were at odds. Probably when we had to buy cookies for a scientist's talk. Part of me understood her reluctance. Harvey was ambitious. But she'd left me alone with this.

And that felt, well, lonely.

"Tusconi," I said aloud. "It's up to you to figure this out."

Pacing, I talked to myself Italian-style, with my hands. Palms up—what to do? How would I proceed? More pacing.

Fist into palm. Got it. Write it out. Make lists. That'll help me think.

I have a large whiteboard on one wall that I use for lists and the like. I grabbed a whiteboard marker and wrote "ideas" and "talk to" at the top. I'd half-filled my whiteboard with a list plus people I might question when I heard a gentle knock. I opened the door, stood aside for Harvey, and shut the door again.

"I feel crummy leaving you with this." Harvey walked over to the whiteboard and stared at it. "Huh. I see what you're up to."

Another knock on the door. This time much louder.

"Mara, are you there?" Seymour's voice. "I want to speak with you."

6

HARVEY AND I FROZE. MY scribbles were clearly visible on the whiteboard, and there was no time to erase them. As department chair, Seymour had keys to all the offices. Harvey pantomimed him putting a key into the lock. My mouth went dry.

We scrambled.

Seymour opened the door and walked in to find Harvey at my desk, earphones on, staring at a computer screen. I sat cross-legged on my yoga mat, eyes closed.

"Why didn't you open the door?" he demanded.

Harvey pulled off her earphones. "What?"

I opened my eyes. "Seymour, what're you doing here?"

"You didn't hear me knock?"

"Guess I was in om-land."

Seymour narrowed his eyes. He waved his hand at the whiteboard. "Rolling buoy, a bunch of names—what's this?"

"Oh, that? Just what happened on *Intrepid*."

"But why would—?"

Harvey interrupted him. "Do you always open office doors when nobody is inside?"

That took him by surprise. Harvey was always so polite.

He pressed his thin lips together and looked sideways at her. "Why do you think I have a passkey? I always knock first, but sometimes I need to check to see if everything's okay. Any more questions?"

We stared at him.

"Good," he said—and marched out.

I let go of the breath I was holding and looked up at

Harvey. The hot pink, dangling earphones and wide eyes were too much.

My chortle morphed into a snort, and in seconds I was doubled up on the yoga mat, laughing like a lunatic.

I managed one "Harvey, I'm so, so sorry" in there somewhere.

Harvey couldn't help herself either. She giggled and held her sides as tears streamed down her cheeks. She pulled a tissue out of the box on my desk, mopped her face, and said, "When he walked in, did you see his face?"

I crawled to the closest chair and climbed onto it. "Like he fell into Alice's rabbit hole." I grabbed a tissue. "But seriously, Harvey, you want to be department head and would be great at it. This is bad for you, and it's my fault."

She shook her head. "I chose to come back."

"But Seymour—"

"Don't worry about Seymour. I can deal with him. If I go for chair and he mentions this, who'd believe him?"

"Guess this shows what I'm doing is risky, at least where he's concerned."

"What *we're* doing, girlfriend. Looks like I'm in it now."

The surge of relief surprised me. I dabbed my eyes once more. "You're one tough babe, Harv."

After Harvey left, I walked over to my window. The floor to ceiling opening—a nod to the 1800s architecture of the original building—allows me full view across Spruce Harbor. A perfect place to muse. The bizarre incident with Seymour had helped me vent some tension. I took in a deep breath. Here I was about to go poking around, looking for answers about Peter's death.

Was my decision to investigate only because, as I said to Harvey, Peter deserved an honest and thorough investigation?

Maybe it was more than that. Last night I stood before another window—Angelo's—facing the ocean, and Peter's death brought me back to a dreadful time.

I was nineteen. In an instant, both my parents had died. After the initial shock, I insisted that the incident was not an accident. My mom and dad would've checked that sub a dozen times and so would their pilot, someone they'd done similar dives with a dozen times. But the authorities wouldn't listen to a grieving daughter.

I couldn't investigate then. But I would now.

The list on my whiteboard was a start. Of the names I'd listed, Ryan was the most obvious person to talk to. But waiting a day was a good idea. Ryan was one of the kindest men I knew, and gossip that he'd intentionally harmed Peter pissed me off. Overwrought with guilt and shame, he was probably in his own hell.

I'd give Ryan a bit of time.

John Hamilton's name was below Ryan's. There were a few reasons why he might be a good person to start with. First, it was odd that he was on *Intrepid* since he wasn't a scientist. Also, it was Seymour's idea for him to come along. Given what Betty told me about Seymour's past, that might mean something—or nothing. And Hamilton was on deck when the buoy fell on Peter. Maybe he saw something. Finally, John Hamilton owned an aquaculture company, something that genuinely interested me. That added up to four reasons to talk with him.

Tapping a pencil against my thigh, I tried to come up with a credible purpose for my visit to Hamilton's facility. Maybe I claim to be curious about aquaculture. That was pretty lame. There must be a better excuse. My brain felt fried, and none came to mind. I needed food—something sweet. And Angelo had asked me to stop by.

I bought a quart of my favorite gelato—strawberry balsamic—and drove to his house. The latest *Spruce Harbor*

Gazette was on the table in Angelo's kitchen. I scanned the front page and waited for his take on the gelato.

Angelo looked at his bowl and wrinkled his nose. "Who puts balsamic vinegar in ice cream?"

The headline piece of the *Gazette* caught my attention, so it took me a few seconds to respond. "It's good. Give it a try."

He sampled a tiny bit. "Huh, you wouldn't expect this to be so tasty."

"Glad you like it. Did you see this?" I pointed to the headline—"Local Seaweed Farm Wins"—and slid the paper over to him.

Angelo dipped into the gelato again and leaned over to read the article. "Could be hype."

"Yeah. Sunnyside Aquaculture successfully develops a super-seaweed. Maine scientists triumph. Etcetera."

Angelo put down his spoon, leaned back in his chair, and crossed his arms. "With all that's going on, why the sudden interest in aquaculture?"

I relayed what Betty had said about MOI, plus my disappointment with the Coast Guard's inquiry.

He frowned. "I've known Betty Buttz for what, forty years? Hate to think of MOI like that. But she's shrewd and might be right."

"Betty suggested I explore on my own. Like she would. I thought I'd poke around, see if anything comes up."

He eyed me. "And poke around means...?"

"I'm trying to figure that out. Basically, I'll look into a few things."

"Like what?"

I explained why it made sense to talk with John Hamilton. "The problem is I need a credible reason to go up there." I glanced down at the *Gazette*. "This aquaculture place. It's John Hamilton's business."

"So?"

"Here's my reason. I'll say I want to know more about the super-seaweed for my Oceanography class. You know, local sustainability ventures. Students love that stuff."

Angelo ran his fingers through his hair. "Seems like a long shot to me, talking to this John Hamilton. What could aquaculture have to do with Peter's death? But I don't see what harm's in it."

I was up at five the next morning and seated at my office desk by seven. It hadn't been the most restful night. Peter and I had discovered the Prospect Institute email only a few hours before the buoy debacle, and that crisis overshadowed everything else. During the night, worries about the email hacking and its implications for my career resurfaced with a vengeance.

I poked my head into the hallway. Ted's door, five offices down, wasn't open. MOI had lured Ted from Duke University because of his pioneering work on ocean acidification and marine organisms. We'd only spoken briefly on the cruise, but he seemed like a good guy and had offered to help me. This was my chance to get to know him.

I returned to my desk and emailed him.

Morning, Ted. Wondering when you'll be in. I need the name of your contact at the Portland Ledger.

Three minutes later, he responded.

Be there by eight.

At 7:55, Ted walked in and placed a cup next to my computer.

"Thanks. What is it?"

"Decaf latte. I emailed you from the Neap Tide and asked Sally what you usually had."

I opened the lid and sipped the milky brew. "Mmm... terrific."

Ted carried a chair over to my desk and sat down. He wore a button-down blue cotton shirt open at the neck with the sleeves rolled up, jeans, and running shoes. With one hand, he pushed dirty blond hair up off his face. His tan set off startlingly blue eyes.

Once more, I envisioned Ted and Harvey as a striking couple.

I ran a finger down my ponytail and got stuck in a tangle halfway down.

"Haven't seen you since we got back," he said. "How're you doing?"

"It's hard. But Harvey and Angelo—he's my godfather—they've been great. How 'bout you?"

"I told my parents what happened—they're down in Boston. They try to be helpful but don't understand." He shrugged.

"If you need to talk, don't hesitate to stop by."

His smile was warm. "I might do that. You want my friend's name. But tell me. The Prospect Institute email. What're you most worried about?"

I swallowed. My mouth tasted metallic and dry. "The other scientists on the list are older and established. It's my credibility as a researcher. You know, respect. Getting grants."

"Colleagues know you haven't cooked the data. But someone might believe it."

"Or use it against me even if they didn't."

"Well, something's just happened I'm guessing will sideline the Prospect Institute. A *much* bigger server hacking. Hundreds of emails between top climate change researchers from the U.S. and Britain."

I felt giddy with relief and guilty that I did. "Damn. I haven't seen the news yet. Still, I'd like to talk with your friend."

"Bob Franklin."

"Thanks. Maybe he could use Prospect's reputation against them. The so-called revelations would backfire. Bob could explain what the out-of-context phrases really mean."

"I like it. And so will he, I suspect." He gave me Bob's email and phone number, then stood. "I don't think anything's out about the Prospect Institute and climate researchers' emails."

"Some buzz on the usual denial blogs, but that's it. So Bob might go for this."

"Good luck. Gotta go. I've got papers to review and cruise data to look at."

I thanked Ted, and he headed to his office. Things were looking up. Ted might become a friend, and the idiots at Prospect might get what they deserved.

I called the *Portland Ledger* and reached Franklin right away. He was excited about my take—how the Institute had turned our scientific conversations into nefarious-sounding smoking guns.

"The editor forwarded the email to me," he said. "I looked at their website. It screams bias against climate change research. So I need to talk to scientists on the list. Really glad you called."

A half-hour later, he had his headline, lead, and story. Instead of exposing climate change researchers as manufac-turers of fiction, the *Ledger* piece would show how Prospect misused our words and phrases to promote their agenda. Done well, the article would educate people about the doubters.

"What about the other papers?" I asked.

"Nothing's come out. I assume other reporters are doing the same thing. Checking the facts."

I hung up, grinning. This was the first good thing to happen in days.

Two items topped my immediate to-do list: contact John Hamilton and work on the cruise data and NOAA proposal. I called Sunnyside to see if I could visit the following day. The aquaculture facility's receptionist said Mr. Hamilton was out on the restricted pier—whatever that meant. Five minutes later he called back. No, tomorrow wouldn't work.

I'd have to drive up there today.

1

B Y MID-MORNING I WAS ON Route 1 North heading to Sunnyside Aquaculture. Maine's coast still clung to winter hibernation, and I zipped past forlorn lobster shacks, ice cream stands, and pottery shops with empty parking lots.

When we spoke on the phone, Hamilton bought my excuse—oceanography students' fascination with Maine aquaculture. Somehow I'd insert questions I really wanted him to address.

South of Winslow Bay, a big sign bordered with intertwined seaweed announced "Sunnyside Aquaculture" in gold letters. A half-mile driveway snaked through spruce forest and ended at an impressively large gray-shingled, two-story building. I pulled open a wooden front door, stepped into the lobby, and gasped. A ring of skylights illuminated a floor tiled in a blue and tan fish motif. Paintings depicting marine scenes hung on cream-colored walls, and an enormous cherry art deco desk gleamed at the far end of the lobby. I turned slowly, taking it in. All this money had to come from somewhere.

The receptionist seated behind the desk stood. "Welcome to Sunnyside, Dr. Tusconi. Mr. Hamilton is waiting for you."

She pointed to a tiled stairway leading to the second floor where John Hamilton greeted me at the top of the stairs. I'd spoken to him for only a moment on the ship and driving up could barely remember a thing about him. Hamilton was shorter than I recalled—hardly more than five feet—with thinning brown hair combed straight back, revealing streaks of pink scalp. As he shook my hand, I examined his face. There was nothing compelling about the man except for his brown, almost black eyes where a spark hinted of intelligence and determination.

I followed John Hamilton into his office. The room was cozier than the reception area.

He pointed to a large picture window. "Aquaculture tanks. You can see 'em from there."

We stood looking out of the window. Hamilton tossed a pencil from one hand to the other. Below, five piers easily fifty feet long stuck out into the water. Each supported two rows of cylindrical tanks in various shades of green, red, and yellow-brown, fifteen or more feet in diameter. Water boiled up inside the tanks. Some spilled over the sides, running onto the pier.

Inside each tank lived millions of cells of a single species of algae—microscopic plants at the bottom of the marine food chain. The algae that grew the fastest, providing the greatest amount of mass in the shortest time, would be the winner. The so-called super-seaweed.

I spotted the likely candidate. On the leftmost pier, four tanks looked remarkably green. Wizard of Oz green. They were a vibrant emerald, almost glowing. Super indeed—and bizarre. I couldn't wait to see the tanks up close.

Beyond the piers, odd-looking white pods bounced in the waves.

"My, an impressive operation," I said.

Hamilton looked down. "Indeed." This was clearly a favorite spot for him.

"Tell me what you do here. The algae you grow, nature of your work. That kind of thing."

He gave the pencil another toss, catching it in his right hand. "Come. Show you the lab."

Back downstairs, Hamilton led me into a bright, well-equipped laboratory. Expensive scientific equipment—large microscopes, instruments for aquatic chemical analyses, an array of glassware—lined three sides of the expansive room.

He directed me to sizeable walk-in incubators on the fourth wall, pulled a handle, and ushered me in.

"Here we experiment with algal cultures."

Glass vats six feet tall sat neatly side by side—some bright red, others shades of green and rusty brown. Banks of bright lights hung over the vats, which bubbled with air delivered by intertwining arrays of plastic tubes.

The room was alive with light, gurgling water, and color.

I whistled. "This is amazing. In this chamber alone, you're growing what, six algal species?"

"More. And there's two other growth chambers."

"John, if you spoke to my oceanography students about growing algae, what would you say?"

He explained the basics of algal aquaculture—that algae needed a good deal of light, the right nutrients, and well-circulated water. I knew all of this but nodded encouragement to be polite.

When he finished, I said, "Ah, this is pretty expensive. Do you have grants? Investors?"

"Both."

I waited for more, but he only led me back into the lab. He gestured toward the lab equipment. "You need no introduction to this."

Hamilton headed toward the door leading to the lobby.

I pointed toward the bay. "But what about the piers? The tanks outside look fascinating."

He stopped and turned around. "Outside? Off limits to visitors now."

"Off-limit *algae*? Why?"

"Let's go back to my office—we can talk there."

I followed him up the wide tile stairs, trying to keep my disappointment and irritation in check. Voicing either would get me nowhere. As I trailed my host, the oddest feeling came over me. Something weird was going on at the facility. I didn't have a clue what was off, but my sixth sense had served me well in the past. I shelved the impression for later.

I returned to the picture window in Hamilton's office with its view of the tanks—the closest view I could get. He gestured to a plush leather chair, and I sat down opposite him. He looked down at his hands.

"Sorry we couldn't go on the piers." He glanced at me. "Frank's worried about security."

"Frank?"

"Frank Lamark. Brains behind the research. Guy's a genius and likes to tinker. Great combination for us."

"So it's important to keep him happy."

He nodded.

"I'm guessing you're experimenting with growing algae for biomass," I said. "Local energy source. So the military, for one, doesn't rely on foreign oil?"

He perked up. "That's right. Air Force especially."

"Tell me more about that. It'll really interest my students."

Hamilton's eyes sparkled. "Algae's a great choice for biomass fuel. The cells grow quickly anywhere there's enough light and water. You don't need expensive farmland."

Again, I nodded encouragement. As he talked, Hamilton's face turned from pale to flushed, and his words spilled out in an excited rush. The man came alive.

"I'm convinced we need alternative sources of fuel for a sustainable future. Algal aquaculture is my contribution to that future. Sustainability, climate change, all that, interests me. That's why Seymour—we've been friends for a long time—why he told me about the *Intrepid* trip."

I blinked. It surprised me that Seymour was anyone's friend.

"I was thrilled to go on your research cruise. Until what happened, of course."

Time to switch gears. "You were on deck when the buoy dropped on Peter?"

Hamilton shook his head. "Lord, yes. I'll never, ever forget."

"I'm just wondering. Did you happen see anything, ah, odd?"

"It happened so fast, you know."

I nodded.

"There was one thing. That photographer."

"Cyril?"

"Never caught his name. He was on deck taking pictures with that big camera. Right before the accident he disappeared. I noticed because a crew member moved over to the spot where he, ah, Cyril, had been."

"Huh." I tried not to sound too interested in this intriguing bit of information.

I stood and looked out at the piers again. I counted four massive cylindrical tanks on each pier—twenty tanks in all. A man on his knees fiddled with a pump beside a tank, his blond hair wet from the squirting water. The whole thing was an impressive, expensive setup. I couldn't imagine what the maintenance alone entailed or cost.

The emerald tanks caught my eye once more.

"Those green tanks at the end. Is that what you're calling the super-seaweed?"

"Yes. The name was Frank's idea. You know, get publicity and more backers."

I pointed to the rows of white pods bobbing a quarter mile beyond the piers. "What are those?"

"Frank calls them flootles. Floating noodles, 'cause they're like pasta tubes. Silly name. Something he's experimenting with—growing algae in floating containers. White to reflect light and lined up in a row so they don't bang into each other when it's rough."

I slid into the leather chair again. I wanted more details about this operation.

"What type of growing media do you use for the algae?"

51

Hamilton shook his head. "Sorry. Don't know that type of thing."

"What about the total number of species you're growing?"

"You'd have to ask Frank."

He couldn't answer my other questions about growing algae either. After all, I decided, he was a businessman and not a scientist.

I stood up to leave. A voice behind me rang out. "John. We go in five minutes."

He leapt to his feet. "Yes, Georgina. Oh, and this is Dr. Mara Tusconi. She's interested in aquaculture. Dr. Tusconi, this is my wife."

Georgina Hamilton towered over her husband. She wore a crisp pink cotton shirt and a pencil skirt. Her red lipstick was a striking contrast to shining black hair that expertly framed her face.

Georgina gave me the once over and extended her hand. "Delighted," she said. "I'll meet you outside, John." She walked briskly out of the room.

I thanked Hamilton for the tour, and he apologized again for its brevity. Back in the lobby, Georgina passed me going the other way. Clicking by in high heels, she gave me a quick nod.

After a bathroom visit, I stepped out the front door of the building just as a black BMW sprayed gravel my way. Georgina sat at the wheel. Her husband sat in the passenger seat beside her, his combed hair barely visible over the headrest. I walked down the wide granite steps. The wind picked up, and bits of sand skittered along the bottom step.

The BMW disappeared from view. I headed for my own car and stopped at an intersecting gravel path skirting the facility. Could I follow it back and get a closer view of the piers? I looked around and—feeling more than a bit guilty—stepped onto the gravel.

The path hugged the building on a windowless side, so I guessed the lab was on the opposite one. I walked quickly. What looked like water flickered through the trees where the path turned behind the building.

I took the turn and stopped. The walkway ended at a cement slab beneath double doors. Two trashcans graced the slab. This was a walkway to a back door. I squinted to catch a glimpse of the piers, but the evergreens were too thick. To see more, I'd have to bushwhack through them, which wasn't a great idea.

I stared at the trashcans. Trash. In the last Miss Marple book I'd read, she poked around in the trash for clues.

I lifted the lid of the first can and peered in. Nada. Must be garbage pick-up day. The other one was most likely empty too, but I gave it a try. Something rested at the bottom. I reached in, grabbed a large crumpled mesh bag, and held it up. The label said "10-10-10."

Fertilizer, probably for the lawns. Certainly not a clue. I threw the bag back into the can and quickly replaced the lid.

I hoofed it back to my car feeling like someone was watching me. The hairs on the back of my neck didn't settle down until I sat behind the wheel. I blew out a breath. Okay, I'd gotten carried away, opening the trash cans. Not nice, but in a list of crimes certainly on the minor side. I looked around. Other than a few cars in the lot, the place was deserted. My guilty conscience must have manufactured a spy.

I pulled out of the driveway and passed the fancy "Sunnyside Aquaculture" sign. John Hamilton's interest in sustainable fuel for the U.S. was obviously genuine. He was truly apologetic that I could not go out on the pier, which he could see I itched to do. All in all, a decent guy.

His wife was a different matter. I'd only seen her for a moment, but the way he jumped when she walked in was curious. I also didn't like the way she gave me the once over—sizing me up, I guessed, to judge who was smarter or maybe more attractive.

Besides that, John had given me a tidbit of information that might turn out to be important. Cyril White disappeared from view right before the buoy fell on Peter.

I stopped to buy yogurt and fruit for lunch before I hid away in my office. Despite my concern about the circumstances around Peter's death, my focus had to be the NOAA grant proposal. The deadline was approaching fast and there was a hell of a lot of work to do.

I logged on to check the streaming data from the buoys we'd deployed. Columns of numbers from the first buoy lined up on my screen—wind direction, wind speed, wave height, water temperature. I looked skyward and sent a prayer to the science angel (there might be one), and ran my finger down the hourly averages.

Damn. The temperature hadn't budged. I rubbed my eyes. Maybe the second buoy would be different. It wasn't, and neither were the others.

I walked to the window and leaned my forehead against the pane. If water temperature didn't increase significantly real soon, the proposal would be much harder to write. I could rely on earlier data, but the argument about rising temperature in Maine's waters wouldn't be nearly as strong.

I was excited about my research and hopeful grant reviewers would be, too. My former grad students and I had evidence that phytoplankton were dwindling in Maine's warmer waters. We'd found declines in numbers and types of phytoplankton, the lovely tiny floating plant-type species that feed Gordy's fish, and speculated the reason might be reduced mixing between increasingly warm—and therefore lighter—water floating on top of denser, colder water below. Phytoplankton need light, and grow in the upper sunlit layer. But as they grow, they use up nutrients. If new water from below can't mix in, the algae starve.

One of my students won a prize for his talk about this research at the last Society of Oceanography meeting. But we'd just started the work, and I needed more grant money to continue.

Thank goodness there was time to write the proposal; I'd taught my half of Introductory Oceanography during the first part of the semester and now another faculty member was in charge. I loved teaching, but the course took a lot of time—hours to prepare lectures delivered electronically to students all over the state three times a week plus more hours to answer students' emailed questions plus exams to grade. Every professor I knew struggled with balancing teaching and research.

My teaching skills hid a secret. Nobody, not even Harvey, knew that public speaking terrified me. Just the thought of it made seasickness seem like a picnic. People from MOI's publicity office asked me to talk to local groups like the Lions and Rotary Clubs, but I made up any excuse I could think of. They stopped asking, and I felt guilty as hell about the whole thing.

By midafternoon, my back hurt from hunching over the computer. I got up to stretch. A heavy knock on my door startled me, and I opened it to a most unwelcome visitor.

"I would like to speak with you, if I may," Seymour said.

Stepping aside, I pointed to a chair and sat back down at my desk, swallowing hard. Whatever Seymour had to say to me couldn't be good. He sat, shifting his weight as if he found the chair uncomfortable. His thin face was impassive, and his steel-gray eyes gave nothing away.

"Well." He cleared his throat. "To the task at hand. I must deny your request for a match for your NOAA proposal."

My heart froze. Matching money from MOI was a prerequisite for even submitting the NOAA grant. "But that means I won't have money for research cruises. I'll miss a whole *year*."

He looked away, then back at me. "Mara, my budget—"

"You've known for months I'd need that match."

"Listen to me. Matching money came from a state program that's been gutted. I just found out about it."

"So nobody in the department will get the match?"

Seymour crossed his legs. "I think I can squeeze out one or two. Your research is, ah, somewhat preliminary. You only have last spring's data. Is that correct?"

"I'll have two years with the new buoy values."

"I believe your proposal is more likely to be funded if you have data from spring, summer, and fall."

Seymour stood, pulled my door open, looked back at me, and left.

I shut the door, returned to the window, and again rested my forehead on the cool pane. My whole body seethed hot with anger. Seymour knew I couldn't collect data without research money. He said his matching budget was slashed, and in the next breath said he'd squeeze out funds. For someone else, of course.

I stepped back and looked out. A breeze picked up on the bay, and several sailors briskly rowed tenders out to their boats for an afternoon jaunt. All smiles in yellow and red wind jackets, they looked so happy. It was, what, only four days since I'd felt that way as I bounded up *Intrepid*'s gangway.

I rolled my shoulders. My back muscles were tense and a dull ache crept up my neck to my head. Time for a break. When I'm low, there's a soul whose gentle ways never fail to lift my spirits. His name is Homer.

I TOOK THE BACK STAIRS down to the basement and pulled open the double doors. The smell of salt-laden air and the deafening roar of seawater jetting through giant pumps greeted me. Throughout the cavernous room, splashing water spilled over aquaria into holding tanks, down pipes, and back to the ocean outside.

I walked to Homer's tank and peered in. He was asleep inside his favorite bottle.

I tapped softly on the glass. "Hi, baby."

Homer stirred, backed out, and touched an antenna against the aquarium window. Without question the two-and-a-half-foot *Homarus americanus* was the prettiest lobster I'd ever seen. His carapace was bluer than most, a striking contrast to the bright red at the tips of his claws.

Homer and I had a human-crustaceal relationship.

Homer's round black eyes followed me as I paced in front of his aquarium. I told him about all my money worries.

"I get that Seymour's in a bad way if his budget was gutted. But he walked in on Harvey and me and saw that list on the whiteboard. I've got to wonder if he denied my match as some kind of message."

I stopped pacing and glanced at Homer. He waved his antennae back and forth.

"Okay, I know that sounds paranoid."

The flowing seawater cadence changed like it does when someone opens the double doors. I turned around, squinted, and scanned the room. It looked empty. Still, I felt uneasy.

Silly. I was just nervous someone would catch me talking to a lobster.

Ten minutes later, I figured Homer had heard enough.

"Hungry, gorgeous?"

I dropped a few mussels into the aquarium. Homer picked them up one by one and used his pincer claw to maneuver mussel meat toward his mouth.

"Homer, what do you think? Is there something rotten in the Republic?"

I dropped in a three-inch mussel.

Before it hit the bottom, Homer snatched the mussel with his big claw and crushed it with one swift snap.

By the time Harvey arrived at our favorite time-out spot on Water Street—the bench beneath a bronze fisherman—I'd calmed down. While I waited for her, I watched kids learn how to sail dinghies around Spruce Harbor and tried to imagine what Peter's wife and children were going through.

No question about it. My problems were minor in comparison.

Harvey had changed into jogging clothes for a late afternoon run. She sat next to me and stretched out her legs. "You okay? You sounded upset on the phone."

"I'll tell you in a minute. First, is Peter's service at nine or ten tomorrow?"

"Ten."

"Okay. I'm planning to visit Sarah in an hour."

"Sarah will be pleased to see you, I'm sure. She's holding up pretty well, considering. So what's going on with you?"

I described Seymour's visit to my office. Harvey twirled a foot and digested what I'd said.

"I'm really sorry, Mara. It's horrible timing for you. But it sounds like Seymour's been blind-sided by budget cuts."

"He was a little nicer than usual. Must feel guilty."

"He's probably got a lot on his plate."

"I guess. But damn it, Harvey, without that grant, I'm absolutely stuck. No money for research cruises, supplies, all of it. And there's that new grad student I told you about. She's expecting support."

Elbows on knees, I leaned over and buried my face in my hands. Harvey let me be until I sat up and stared out across the harbor. She put her hand on my arm.

"Couple of things," she said. "First, you're one of the most creative scientists I know. And smart. Plus you work like a dog."

I managed a small smile. "You don't like dogs, Harvey."

"Listen to this. Biological Oceanography at NSF just announced a new program that's right up your alley. Check your email. Something about ecologists working with people impacted by climate change. Fishermen, city managers, that kind of thing."

I jumped up. The National Science Foundation was a great funding source. "Really? Gordy and I were just talking about high temperature and fish."

"Also, proposals are due really soon, so there'll be less competition than usual. I'm guessing money came down from the hill. Maybe you can use some of your NOAA ideas for this one?"

"I sure could. And there's no match required for an NSF proposal."

"Right, and thank goodness for that. A match for my new NSF grant would've been enormous."

I laughed, and boy it felt great. "What'd you get? Half a million for your ocean acidification project? You're amazing, Harvey."

"Thanks." She glanced up the street. "Um, I've got to get going soon. You drove up to Sunnyside this morning? Did you learn anything?"

I described the visit in sequence—from my amazement at the expensive facilities to Georgina's unpleasant behavior.

"Curious the piers are restricted," Harvey said. "Maybe they're waiting on a patent."

"Now I'm even more curious about the so-called super-seaweed. I saw the likely candidate from a distance. It looked like something out of *The Wizard of Oz*. Emerald green."

Harvey stood and stretched a leg on the bench. "You could ask Ted if he knows what's going on up there. Algae are his specialty, after all."

I eyed her to see if maybe she'd been chatting with Ted herself. But she was all innocence as she switched legs.

"Good idea. I'll also check out Georgina Hamilton on the internet. See what her background is."

"Okay. Time for my run. Call me tonight if you find out anything interesting."

She took off so quickly, I thought she didn't hear my "Hey, thanks, Harvey." But she waved both hands high in the air before she disappeared around a corner.

In running shorts and a T-shirt, Ted passed me on my way back to MOI. I smiled as he zipped by. Huh. Maybe he and Harvey had a jogging date.

I drove to the Riley's cottage in town, parked in front, and grabbed a jar of my homemade blueberry preserves from the passenger seat. The clock on the dash said four-thirty. Not too close to dinnertime for my visit. Crushed shells crunched beneath my feet as I walked along the path to the front door. Last fall, Peter told me how he'd added the shells, proud of his workmanship. I rang the front door bell and waited as memories of happier times washed over me.

A small, plump woman with blond, blunt-cut hair and an air of efficiency swung open the door. "May I help you?"

"I'm Mara Tusconi, one of Peter's colleagues. Is Sarah here? I'd like to see her—just for a few minutes."

The woman stepped aside. "She is, dear. I'm Katherine, her mother. Sarah's in the kitchen."

Katherine led the way and returned to the task I'd obviously interrupted—chopping veggies.

Sarah rose to greet me from a chair at the kitchen table. Tiny, fair-haired, and pale even in summer, now she looked wan and even more diminished in her grief. She extended her hands toward me. "Mara. How lovely for you to come."

I set the preserves down on the table, then held both her hands in mine. "Sarah, Peter was a good friend, important to all of us. If there is anything I can do—"

She squeezed my hands and released them. "Thank you for the visit and preserves. Come sit. Coffee?"

Katherine placed a steaming mug on the table before I could reply. I pulled out the chair next to Sarah's and sipped the strong brew as she talked about the twins and the next morning's memorial service. I described a few projects and students Peter and I shared.

"He was terrific to work with, Sarah. The best. I learned a lot from him. We had some good laughs."

She sniffed, pulled a tissue from the box on the table, and dabbed the corners of her eyes. "Thanks. Those are wonderful memories." She looked down at her hands. "Tell me what you saw, Mara. I know a little. I think Peter took your place?"

I'd anticipated Sarah's questions and considered what to say—and not say—on my drive over. I kept to the facts— why Peter took my place, that the first deployment Harvey supervised went well, and that something about the buoy concerned Peter. I didn't mention the thoroughness of MOI's investigation.

"It happened so fast, Sarah. I wish I could tell you more."

She nodded and teared up again. We sat in silence. She said, "Something bothered Peter right before he left. I don't want to trouble you—"

"Sarah, what is it?"

She glanced to the side. "I don't know. It's probably not important. You're so busy."

I put my hand on her arm. "Please, Sarah. I want to help."

She searched my eyes. The pain in hers was heartbreaking.

"It's an email. He talked about it right before he left. He was angry."

"Really? I don't think I ever saw Peter angry."

"It took an awful lot to upset him. That's why I've been wondering about it."

"Do you have the email?"

"Be back in a minute. I printed it out."

I only managed a few more sips before Sarah returned. The message she handed to me read:

This one's close to home. A bioengineering claim that seems dubious. It's not my field, but this scientist hasn't published much. I called to arrange a visit after I get back.

I frowned. "I don't know what to make of this. Who is Peter writing to?"

"He was part of an internet group called 'sci-fraud', I think. They're watchdogs for scientific dishonesty in papers that get published. He's writing to those people. That's all I can tell you."

I sat up taller in my chair. "Really? I didn't know he did that."

"He said something was bogus and was angry, like I said."

I scanned the message. "I'll work on this, Sarah. If anything comes up, I *promise* to get right back to you."

She dabbed her eyes again. "I'd very much appreciate that."

We talked for a few more minutes. I confirmed I'd see Sarah and her mother at the memorial service and left.

After a quick stop for groceries, I drove the bumpy mile down to my house. The road winds through stands of spruce so tall and dense, a carpet of bright green moss is the only thing that grows on the forest floor. Pencil-straight spruce, damp pillows of moss, gray rocks scattered by the last glacier—it's my own bit of Maine rainforest.

My house stands alone at the end of the road where the woods open up to reveal the rocky Maine shore. I bought the run-down cape five years earlier, and with a lot of help from Angelo turned it into a classic gray-shingled treasure. It sits thirty feet up off the water at high tide. Twin sets of oversized windows look down on huge granite boulders, a cobble beach, and the Atlantic beyond. My little home is my joy and haven.

Glass of wine in hand, I grabbed a fleece throw and settled into my Adirondack chair on the bit of grass out front. The sun had already set and a chilly breeze ruffled the water. I snugged the throw tight around me.

I truly didn't know what to make of Peter's message. A dubious bioengineering claim could be something biomedical, agricultural, genetic. To find out, I needed to contact whoever was in charge of this "sci-fraud" group. But surely their exchanges were privileged. That made finding these people a challenge.

Peter's uncharacteristic anger was curious. I could not begin to imagine how the email and his reaction to it were connected to what happened to him on the ship. But if there was a connection—and someone tampered with the winch—maybe Peter was the intended victim after all.

Lots of questions, no answers. Sarah wanted answers, and I was going to do whatever it took to find them for her.

Above the water a tern swooped, circled, swooped again. It dropped straight and fast into the sea to grab a fish.

My computer sat on the kitchen table. I flipped it open and scanned the National Science Foundation website for the program Harvey mentioned. I entered "climate change" and "collaboration" as search terms and found a new initiative called "Adapting to Climate Change"—ACC. Harvey was right. The deadline was very soon, and I could easily use prose from my NOAA draft. I needed to talk with Gordy to see if he was interested in teaming up with me.

The microwave clock read six-thirty. Gordy had the evening meal he called "suppa" at five. I called him up and briefly described the NSF program, and explained that we'd have to apply the latest climate change research to fishing methods.

"Would you like to work with me on a project we'd design together?"

"You're kiddin', right? Eggheads and real folk?"

"That's the point. Science that can be used in the real world. It's why they developed the program."

"I dunno. This global warming thing. You know I think it's cr— er, isn't happenin'."

"I do. Again, it's part of the challenge."

"You're my cousin and all, Doc. But I don't know."

"Gordy, if we get the grant you'll get a very good salary. And so will the other fishermen."

"Now *that's* a diff'rent story. Let's give it a go. I'll be at that office o' yours seven sharp. In the mornin'."

I grinned and agreed. Gordy was up well before five, so anyone who rose after that was "sleepin' in."

Back at the computer, I searched for "Georgina Hamilton" and quickly found her. Georgina had impressive credentials. She'd received her law degree from Harvard, worked for what looked like a prestigious law firm, and served on the board of several companies and foundations. I recognized them all except the last one listed. I entered the name into my search engine and was surprised to learn that an oil conglomerate I'd never heard of was among the richest corporations in the world.

Georgina probably earned a bundle, which explained the fancy car. Undoubtedly, she knew lots of rich people as well. Maybe some of the money for Sunnyside came from them.

I ate warmed up lasagna for dinner and called Harvey.

"How was your visit with Sarah?" she asked.

"Like you said, she's doing okay, considering. It's incredibly sad. When I told her a few stories about Peter, she cried."

"She'll treasure those memories."

"Yeah. And there's something else. Sarah showed me an email message Peter wrote to some sort of sci-fraud group. Know about them?"

"Don't think so. What did it say?"

"Something about a local scientist and bogus bioengineering claim."

"That's certainly curious."

"The email's bothering Sarah, partly because Peter was so angry about it. She also said it was one of the last things he talked about before he left for the cruise."

"She'd have some closure if you can learn what he was talking about."

"I'm going to help Sarah if it kills me."

The next morning, Gordy paced around my office like a fish circling an aquarium. He'd changed into his spring-summer-fall onshore attire—tan canvas shorts fringed at the hemline, brown ankle-high leather boots with white socks, white T-shirt. He chewed on an unlit cigar.

"Tell me exactly what they want so we get the money."

I read from NSF's description of the program. "Um, proposals will be judged on the likelihood that collaboration will have significant impacts on the target community."

"So, in English that means we gotta do something that really matters to Maine fishermen?"

"Yes. You'd take what you learn from scientists and modify what you do. Like where or how you fish and what you fish for."

Gordy shook his head. "We're talkin' about guys who think this warming stuff's a bunch of lib'ral crap. You're askin' too much."

"You've told me that before. We have to figure out what'll work and matter. *And* it has to go two ways. Researchers must learn from fishermen. Like the type of data you need in order to make better decisions."

He smirked. "My buddies'll love that. Teachin' eggheads for a change."

Coming up with a workable idea was much harder than I imagined. To every suggestion, Gordy said "nah," "nope," or "you jokin'?" I suggested that the fishermen make water temperature graphs. The response: "you're friggin' kiddin' me."

Forty-five minutes later, I threw out an idea he didn't immediately reject.

"What if we used your boats as research platforms? I mean, as the way to get information on fish density, water temperature, that kind of thing."

"Huh," he said. "Now that might work. If you want to hear swearin', jus' ask these guys how fisheries biologists get numbers on how many fish are where. This'd give us a chance to show we're right."

"You know, Gordy, I've never been on your lobster boat. What's she like?"

"*Bulldog*'s an old lady. Moored in the harbor. Dark blue, Cape Island style. She's got brand new railings, but I gotta fix those rusty scuppers."

"I'll look for her and work with our idea. See if anyone else has tried it, and get back to you in a few days."

He grabbed his baseball cap from my desk and settled it on his head. "Fishermen's League meets ta'morra night.

I'll pass this by them. If I pitch it right, maybe they'll bite."

He headed for the door.

"And Gordy, in the proposal we'll have to use the term 'fishers' instead of 'fishermen.'"

He pulled the cigar out of his mouth and looked back at me. "Gals are gonna be part of this. Some are damn good fishermen." He winked and shut the door behind him.

9

THE SPRUCE HARBOR CHAPEL, A classic New England Congregational-style church, sits on a rise overlooking the harbor. Scientists, students, MOI workers, plus Peter's family and friends packed its pews.

I slipped in just as the organ music stopped and the minister called us together to remember Peter Riley. From my seat in a back row pew, I could barely see Peter's family—Sarah and the twins with people I assumed were his siblings and parents.

My god, the twins. Sarah would have to raise them on her own. I had no idea if Peter left her any money or had insurance.

I glanced around. Most of us from *Intrepid* were there. Ted sat three rows in front of me, his blond hair neatly combed. Harvey's full-length black coat poked out from the end of a pew on the other side of the aisle. Ryan, his face drawn and pale, leaned over the rail of the chapel's balcony along with a few other crewmembers. Seymour must've been in the crowd somewhere.

The grief in the crowded room was palpable. I held back my tears until my throat was tight but finally let them fall. One of Peter's brothers did his best to recall funny incidents from their childhood, but he broke down and cried, unable to finish. After the service, little groups stood on the lawn outside the chapel, sharing hugs and wiping tears.

I found Harvey. We hung on to each other until she stepped back and searched my face. "You okay?"

I shrugged and dabbed my eyes with a soaked tissue. "I keep thinking that this never, never should have happened.

For Peter, for Sarah, we have to get to the bottom of this."

"We'll work on that. And Mara, I kept thinking this could be your funeral."

The ground rocked a bit. I linked my arm in Harvey's. "Let's stand in line for Sarah."

We reached her and the rest of Peter's family and murmured condolences. Katherine stood beside Sarah, who tried hard to be gracious but looked ghostly and on the edge of collapse.

I drove out of the parking lot utterly drained—but with renewed determination to find out why this family lost a son, a brother, a husband, a father. Peter was a gem, so loved. If his death wasn't an accident, the guilty person simply had to be exposed. And maybe I could help.

I might also learn if the out of control buoy was meant for me.

Sunnyside Aquaculture was my first step in that direction. I'd left the place with questions. To get some answers, I needed to talk with Ted.

On my way to MOI, I picked up the *Portland Ledger*. If Bob's piece had come out, Ted would want to read it. In the car I flipped through the paper. There it was —"Climate Change Deniers Get It Wrong." I skimmed the article and closed the paper with relief.

I reached the landing. Great—Ted's door stood open. I stopped in the bathroom and redid my barrettes. I recently had a trim, so my hair fell evenly over my shoulders and down my back. The auburn color set off the rich brown of my go-to-church coat. Nobody had to know about the slightly ratty cotton turtleneck under my sweater.

Even though we're on the same floor, Ted had traveled a lot since MOI hired him, and I'd never been inside his office. I stood on his threshold and quickly took in the detritus of a busy professor. On the floor, a chair, and his desk were piles of what looked like student theses and journal articles.

A bike helmet, running shoes, and other sports paraphernalia were toppled in the corner by the window.

His back to me, Ted stood at his window, still in his shirt and tie.

"Knock, knock?"

Ted turned. His smile, thin at first, broadened as he blinked and appeared to come back to the present. He'd loosened his tie, and a suit jacket hung over his desk chair.

"Mara. Sorry. Just thinking about Peter and his family."

"May I come in?"

"Sure. Ah, let me clear a chair for you."

He moved one of the smaller piles from one chair to another. I sat down and put the newspaper on my lap. "It was a lovely service."

Ted took his own seat, wriggled his tie loose, and pulled it off. "That's better. Nicest and saddest memorial service I've ever been to. And the chapel's a beauty. Being new, I didn't know Peter long. But he was a pleasure to work with and very smart."

"Everyone loved Peter."

"It's different for us," Ted said. "The ones who saw it."

No need to ask what "it" was.

I nodded. "That scene—it plays like a movie in my head. When I least expect it, you know."

"Yeah. I've had some bad dreams."

"Look, Ted. I can come back another time."

He ran his fingers through disheveled hair and leaned back. "I'd rather you stayed. Did you stop by just to chat?"

"Um, no. Couple of things." I handed Ted the newspaper. "Here's the first one. Franklin's article is on page five."

Ted spread the paper out on his desk. His hair fell forward as he pored over the article. "Wow. Bob did a great job. Listen to this—

Communications between climate change experts must be viewed in context. The Prospect Institute screamed about the word "trick." But in their email exchanges, two scientists used "trick" only when they discussed ways to compare temperature data calculated from tree rings with actual thermometer measurements. As you can imagine, that's "tricky" to do.

"Bob talked to a couple of scientists—he refers to you, Mara—and it looks like he quoted them correctly. You must be pleased."

"And relieved. Thanks again for recommending him."

"There was something else?"

"Yeah. I visited Sunnyside Aquaculture and have a couple of questions. If you're up for it."

He raised his eyebrows. "You're kidding. Why'd you go up there?"

Right. Ted didn't know what I was up to. My cheeks turned hot.

"Have you met Betty Buttz?"

"I know her by reputation. But I've only seen her zip by."

"I ran into her a few days ago, and, well, she warned me that MOI might limit their investigation. To protect the institution's reputation."

"Huh. I suppose she knows the place better than anyone."

"Betty said I should poke around myself. You know, to really find out why that buoy dropped. I owe it to Peter, especially since he took my place. The Coast Guard officer, what was his name? Wilson? In the interview, he called it an 'accident.'" I air-quoted the word. "That pissed me off."

"He talked to me, too, and the interview was pretty short. But Sunnyside? I don't get *that* at all."

Ted sounded annoyed, but maybe he was just perplexed.

I hurried to explain. "Couple of reasons. John Hamilton—he's the owner—was on deck when the buoy fell on Peter. Maybe he saw something. Also, it was odd he came along, and I wanted to find out why. And I'm interested in local aquaculture and figured at least I'd get information for my oceanography class."

"Guess that makes sense."

"So do you know anything about Sunnyside Aquaculture? Their focus? Where they get their money?"

"It's local fuel research. Growing algae for biomass. Actually, a couple of weeks ago some investors from Boston drove up to talk to me about it."

I wasn't surprised these people sought Ted out. Growing algae to replace petroleum fuels was a hot idea, and Ted had a national reputation in the field of phycology—the study of algae. His lab was filled with flasks and bottles holding liquid colored all shades of green plus red, pink, yellow, and brown— different types of algae growing in culture. He'd shifted his research focus to effects of marine acidification—more acid ocean water caused by higher levels of carbon dioxide—on plankton. But if you had trouble cultivating a particular algae, Ted was the one to talk to.

"This is related to the super-seaweed?"

He shook his head. "Calling it 'super' is overdoing it. But Sunnyside is trying to attract more funding."

There'd been a salmon aquaculture farm north of Spruce Harbor in Winslow Bay where fish were grown in floating pens, Ted told me. The owners lied about the number of fish and miscalculated the daily flushing rate of the bay's water. As a result, nutrients from fish waste turned it into a green, soupy mess of seaweed. I knew the story and had used it in an oceanography class to show why aquaculture, contrary to the opinion of some, is not the simple answer to overfishing.

"It's the perfect spot for Sunnyside," Ted said. "Good access to seawater, lots of sun, buildings in good shape."

Ted went on to tell me about the different types of algae Sunnyside was trying to grow. "You'll be especially interested in *Chloronella*. That's the algae featured in the *Gazette* piece."

As a biologist, Ted referred to the organism as algae, as opposed to seaweed. I generally consider "algae" microscopic and "seaweeds" things like kelp—big plants left behind on the high tide line at the beach. But others use the terms interchangeably.

"But what's all this about? Why call it super-seaweed?"

Ted looked at me intently. Ted looked at me intently and appeared to make a quick decision.

"If I tell you this, you have to promise not to discuss it with anyone."

Naturally, I said I'd keep such a promise—and was secretly pleased he'd share this secret, whatever it was, with me.

"They've used genetic engineering to add the nif gene to the algae's DNA."

I stared at him open-mouthed. This was certainly an unexpected achievement. Marine biologists, including Ted, had been trying hard for a long time to genetically engineer algae to make their own amino acids and proteins—as bacteria living in the roots of beans can do. It would save an enormous amount of money because commercially made nitrogen—what you get in bags of fertilizer—is extremely expensive. With the nif gene added to the algae's DNA, you wouldn't need to feed the algae commercial nitrogen because the cells could use nitrogen gas naturally dissolved in water and make their own proteins.

It sounded like Sunnyside had indeed made a super algae.

"Wow. That's incredible. The biofuel industry has been at this for years, pouring money into the research. The military too—to get us off foreign oil. And we have people right up the road who've done it?"

He nodded.

"But how do you know all this?" I said, a little too abruptly.

Again, Ted studied me closely. I felt a twinge for my rudeness.

He said, "Frank, Frank Lamark, works at the algae farm, as he calls it. We grew up together, and he followed me when I moved to Maine. He keeps me filled in, at least as much as he can. They have to be very careful before they patent what they are doing, but it's about to happen. Frank's really proud of himself because he's responsible for the genetic engineering work. We're about the same age and have always competed with each other. You know, who was the better jock, who went to the better college and grad school. Looks like he one-upped me this time."

"You don't look upset."

"Nah. Frank's always felt inferior. To tell you the truth, he has a chip on his shoulder. So this is a good thing. Maybe he can feel proud about what he's achieved and move on. I do care about the guy. He feels like a brother. Growing up, people thought we were. I guess that's because we spent so much time together."

"Frank and the super-seaweed," I said. "Um, is it okay if I share this with Harvey? We're working together on all this. Harvey won't say a word."

"Of course she won't."

"I'm curious about John Hamilton. Can you tell me anything about him?"

Ted shook his head. "First time I met him was on the cruise."

There was one more thing I wanted to ask. But I hardly knew Ted and wasn't sure it was appropriate. I stood and made a quick decision.

"Ah, I have one more question."

"What's that?"

"Do *you* think what happened to Peter was an accident?"

Ted frowned. "Well, the captain and Seymour, they're both calling it that." He looked at his watch. "My parents will be here in an hour. There's a couple of things I need to do."

I headed for the door. "Thanks so much. You've been a huge help."

Ted nodded goodbye.

I walked right into Seymour, who was standing a few feet down the hallway. How long had he been there?

Seymour bent his balding head and looked down at me, his glasses balanced on a ski jump thin nose. "Don't tell me you think what happened to Peter wasn't an accident."

I lifted my chin. "The safety standards for those winches are top of the line."

"So now you're an engineer? You've got research to attend to, Mara. I suggest you focus on that and not what you know nothing about."

I brushed past Seymour, marched to my office, and shut the door. I simply had to leave his nastiness behind. To get my mind off him, I started to look at the buoy water temperature again.

My hand hovered over the keyboard. Peter's memorial service was just a few hours ago, for God's sake. And here I was fretting about temperature data. Compared to Sarah and her family, my angst was trivial. I could put it aside, at least for a little while.

"Okay," I said aloud. "No data checking—for a week, at least. That's an order."

The grant proposal took up the rest of the afternoon. I carefully read through the Adapting to Climate Change description. This would be by far the most difficult proposal I'd ever written.

Most oceanographers I knew were trained in basic, not applied, research. That included me. We founded our studies on previous work and fundamental theories. But applying our knowledge to real-life issues like how Maine fishermen could change their methods as oceans warmed? That was, as Gordy would say, a "whole 'notha thing."

By late afternoon, I was bushed and needed a break. I drove out to the coastal preserve on the edge of town—a great place to jog, which I try to do at least three times a week. Along the winding path through the spruce-fir forest in the spring, I usually inspect swollen tree buds or bend over to examine sand strewn across the path, evidence of storms. But I barely noticed where the trail was taking me as I mulled over my visit to Sunnyside and conversations with Sarah and Ted.

Jogging along, I talked to myself, gesticulating like a good Italian. Half-Italian, half-Irish, to be precise.

Starting with Sunnyside, I tried to organize and connect what I'd learned over the last few days. The aquaculture facility was costly to set up and operate. Ted mentioned investors, so that must be where the money came from.

Being denied access to the piers was a surprise, but given what I now knew about Frank's genetic engineering research, understandable. In all likelihood, the emerald green tanks at the end of the pier held *Chloronella*, the genetically altered super-seaweed.

Genetic engineering. The email message Peter wrote referred to bioengineering and someone who hadn't published much. Maybe Peter was referring to Frank.

I touched the side of my head with my forefinger—the Italian "are you nuts" gesture. Too much of a coincidence.

I dodged a tree branch, thinking of others I'd spoken with. John Hamilton was a decent guy devoted to the algae work, and he apologized about restricted access to the piers. Very different from his wife, who I guessed was ambitious and competitive.

Then there was Ted. I was flattered he confided in me about the *Chloronella* research. But in retrospect it was odd that he shared privileged information. Unless I read him wrong, he also was startled about my trip up to Sunnyside.

Besides that, the whole genetic engineering business bugged me. Frank claimed success with a problem that eluded scientists with a good deal more experience. Maybe he was lying.

My skepticism led to another conundrum. If Frank never created the super-seaweed, it made no sense for him to broadcast that he had.

Pausing for a moment, I said aloud, "It doesn't make sense."

At the outer loop of the path, the boardwalk spur took me down to the beach. Out there in the open, a cold wind whipped up sand, stinging my bare legs. The whole mess whipping around my brain was driving me nuts. I headed back to the protection of the trees and took the advice of one of my favorite detectives—the brilliant Hercule Poirot.

Just let my "little gray cells" do their thing.

Back in the parking lot, I stretched against the passenger side of my car and stepped down in a low runner's lunge. I heard a vehicle pull into the adjacent spot. Cyril White, the photographer on *Intrepid*, climbed out of a spanking new green truck as I straightened up. His chocolate-colored eyes were huge.

"Sorry, Cy. I didn't mean to surprise you."

He stared, then pulled himself together. "Dr. Tusconi. Ah, um. I didn't see you."

"Of course. I startled you. So how are you?"

"Me? Okay, I guess."

"And how did your photos of the cruise turn out? Before what happened, of course."

"My photos? Fine."

"Could I come by and look through them and the video sometime? I'm, ah, trying to get a better idea about what happened to Peter."

"Right. Sure. Hey, I'll take off now." He skittered down the trail.

I stared after him. Cy was so friendly on the ship. Why was he acting jittery now? I shrugged. My experience with students had taught me that the life of a young man just out of college is sometimes a mystery.

I pulled up to Harvey's house in time for dinner. Talking through anything over food is in my blood.

Harvey lives only blocks from MOI. She usually walks to work and leaves her truck in the driveway. Harvey had a silver spoon childhood, so you might expect her to drive a Lexus or BMW. Instead she's got a full sized Dodge Ram. She uses it to go off road and hunt deer, moose, even bear. It's one reason why she moved to Maine.

Deer and moose I get, but not bear. I think she's nuts.

Inside Harvey's neat-as-a-pin ranch is the most organized kitchen I've ever seen. Behind clear cabinet doors, plates and glassware are arranged in horizontal and vertical perfection—and her marble counter tops are completely free of any clutter.

I perched on a kitchen barstool with the inevitable glass of wine. "Thanks for dinner at such short notice. I need some help piecing things together."

Harvey stood in front of her open refrigerator, two plastic bags in her hands. "Sure. Chat with me while I chop."

"You know how we feel at the beginning of a new research project? We look at the early data and can't make sense of the separate pieces?"

"Yes. It's extremely frustrating."

"Well, that's where I am now with all of this."

I sipped my wine while Harvey chopped tomatoes and peppers. As usual, she looked terrific in gray linen pants, matching sweater, and black flats. I glanced down at my old brown corduroy jeans and faded orange turtleneck and sighed.

Harvey finished, poured herself some wine, and took the stool next to me. "Go ahead."

I described what I learned from Ted about Sunnyside, including Frank's success with the nif gene. "This bioengineering business. It doesn't make sense that high-powered labs have been working on the problem for years and a biologist we've never heard of succeeds."

"But what about the tanks you thought contained *Chloronella*? They looked emerald green, like they had plenty of nitrogen."

"Right. Puzzle number one. Then there's the email Sarah showed me. It's tempting to think the fraudulent scientist was Frank, except most biology labs these days do genetic engineering work of some sort. Peter could've been talking about anyone."

"But when you called last night, I'm pretty sure you said Peter mentioned a local scientist. That narrows the field a lot. Besides MOI, what other local biological research labs are there?"

I stared at the ceiling. "Can't think of any."

"We've hired that new microbiologist. She's on my floor. I can ask. But I'm pretty sure she studies deep sea vent bacteria, nothing to do with bioengineering."

"Go ahead. But I still think it's a coincidence. Local's a pretty vague word. Peter could've meant New England, for all we know."

"Come on, Mara. Let loose for a minute. Go with the idea Peter *was* talking about Sunnyside. See where that takes us."

Harvey drank some wine, rubbed her nose, sipped more wine. "What I can't get past is why Frank would claim he created this super-seaweed if he hasn't. If this is biofuel research, the fraud'd be obvious when someone else tried to grow it."

"I know. But if it's a fraud, again, why are the *Chloronella* tanks so ungodly green?"

"What if," she said, "Frank thought the bioengineering *did* work in the beginning but later realized it didn't. He would've attracted investors because genetic engineering is sexy."

"Okay. I'll play your game. But suddenly, the algae doesn't grow so well. It's not at all 'super.' Frank panics and restricts access to the piers."

Harvey took over. "But possible investors want to see the super-seaweed. Frank needs money, so to buy time, he's got to impress them. The algae has to be really green and healthy." She frowned. "But I don't know how he'd do that."

The image popped into my mind. A bag labeled 10-10-10.

"Wait a minute, Harvey. At Sunnyside, I looked into a trashcan and saw a bag of fertilizer—10-10-10. Like for lawns. But there aren't any lawns there. The landscaping's all stone."

Harvey raised an eyebrow. "You looked in a *trashcan?*"

I shrugged. "I know. I felt guilty as hell. But why was the bag in there?"

We looked at each other and said it in unison. "*Because Frank's adding fertilizer to the Chloronella tanks.*"

10

I JUMPED UP AND MARCHED around the kitchen. "It'd be so easy. Frank could buy artificial fertilizer anywhere. All he'd have to do is dissolve it in seawater to make a nitrogen-enriched concentrate and add it to each tank. Sometime when his technicians aren't around, like at night." I leaned against the refrigerator and crossed my arms.

"And," Harvey added, "*Chloronella* would take up the nitrogen, grow like crazy, and look as green as a golf course. But assuming that's what Frank is doing, for whatever reason, his hoax would come out soon enough. That's still what I don't get."

"Let's talk about this later. I'm starving, Harv."

We ate in Harvey's dining room. She'd inherited her parents' furniture, pieces that would probably excite folks on *Antiques Roadshow*.

We chatted about a wild turkey hunting trip she might take.

I speared the last shrimp with my fork and held it up. "Great shrimp scampi, Harvey."

She just looked at me.

"You've been pretty quiet the last couple of minutes," I said.

She slid her plate to one side and leaned forward on her elbows. "Just wondering where you go with this. Even if Frank's fertilizing the tanks, the piers are off-limits, so there's no way to find out."

I sighed. "Yeah. Maddening."

"Is there anything else? I mean, from your trip up there?"

"John Hamilton's totally into his algae-biomass work.

The guy's a milk-toast but a different person when he talks about that."

"And he trusts Frank?"

"Loves the guy. He leaves details about growing algae to Frank. So if Frank's fertilizing those tanks, Hamilton wouldn't know." I shook my head. "Boy, that'd be a shocker for him."

"What about your talks with Sarah and Ted?"

"I asked Ted if the buoy was an accident. All of the sudden he said he had to leave. To be with his parents. But the worst part is that Seymour was standing outside Ted's office. He overheard and told me to mind my own business."

"Hmm," she said. "Ted's mother is a stickler about time. He probably was just anxious to get going."

"You know Ted's mom?"

Harvey blinked. "I meant *if* she's a stickler about time."

"Oh."

She spoke quickly. "And it sounds like Seymour wanted to talk with Ted and stood outside the office until you finished your conversation."

"He always appears out of nowhere at precisely the most awkward times."

Harvey stacked our dishes and carried them into the kitchen. I collected glasses and followed her.

She rinsed dishes in the sink. "What about that photographer's photos? He was on deck with the rest of us when it happened."

I handed her the glasses. "I saw Cyril, out at the preserve today. I'd just come back from a run. He acted weird. Not friendly, like he was on the ship. Huh—wait a minute."

She turned off the water. "What?"

"Something John Hamilton said. That Cy stood in the same place taking photos until right before Peter went down."

"Hard to know what to think about that."

"I agree. Okay, where are we? I talked to John Hamilton. Who's next?"

"Ryan's the obvious next person, Mara."

Harvey was right, of course. But Ryan was a good friend, and I'd tried to protect him. It was time to let that go. "He is. But I can't walk up to Ryan and say 'Hey, why did you let that buoy drop on Peter?' How do I approach him?"

"Do you know any of Ryan's friends? Someone you trust?"

"I do. Gordy knows Ryan and so does Connor. Irish chums, you know."

"Connor?"

"He's Angelo's good buddy. They fish together."

Since Connor Doyle spends most of his time on or around fishing boats, just after sun-up the next morning I walked Spruce Harbor's waterfront and looked for him. Hand shielding my eyes from sun reflecting off the water, I peered through spaces between wooden houses circling the harbor. Painted white, rust, and dark green, they were wavy versions of themselves in the shallows. I passed piers smelling of dead fish and littered with lobster traps. The tide had turned and boats with names like *Little Lady* and *Sea Goddess* swung on their moorings.

Connor, who can list the ships that had carried his Irish ancestors to Ellis Island, moved to Spruce Harbor when he retired early from the Augusta police force. I'd asked a few times why he left the force but never got a clear answer.

Connor says he chose Spruce Harbor because he likes the mix of fishermen, MOI workers, scientists, and summer people. Our town shares a history with many others on the Maine coast. There's a working harbor, with lobstering the mainstay fishery. Now, alongside the fishermen are people "from away" who bought rundown cottages and fixed them up with decks and garages. The collapse of the cod fishery was only one assault to a declining fishing tradition. As a result, away-people outnumber ones who make their living on the water.

I jogged toward a square, solid figure with curly black hair on the beach at the south end of town. Connor swore like a sailor as he splashed around in his knee-high black boots, trying to haul up his twenty-foot fishing dory. She's a beauty—green on the outside with shining mahogany trim and a cream interior.

I took the bow and helped him maneuver the boat over some rocks and onto the beach. "Why not use the town dock?"

Connor tipped his head back toward the water. With his black ringlets, ruddy cheeks, and blue eyes, he looked like an altar boy who suddenly turned fifty. "I've been tryin' to find first blues of the year since before dawn and got to have some coffee. Headin' for the Neap Tide."

Connor's "ye-ah" and "hahve" were true Maine.

"I'll walk you up. You can finish telling me about the west coast of Ireland."

We started the climb up Water Street. "Your great grandma was born in County Kerry. Savage country."

"Is that bad or good?"

"The best. I still dream 'bout the mist on mountain tops 'n meadows so green it hurts the eyes. And below, ocean crashin' on those rocks."

"My father's side also came from the sea," I said. "Naples."

"So 'course you've got extra salt in the blood."

At the Neap Tide, Connor ordered a coffee to go and teased owner Sally until she blushed, which I'd never seen her do. On our way down Water Street, he popped the lid and sighed. "Gaud, Sally makes the best brew."

Because it's steep, Water Street offers a great view of Spruce Harbor. As Connor sipped his coffee, the morning sun turned a half dozen lobster boats gold as they headed out to sea.

Connor interrupted my reverie. "Speaking of Ireland, Mara, how 'bout meeting some Irish lads? I know a few you'd probably go for."

"Thanks, but as I've said before, I like my independence just fine." We'd reached the bottom of the hill, and to change the subject I added, "Connor, I need your help."

He stopped short. "'Course."

"It's about last week on *Intrepid*." I looked around and motioned to a flat-topped boulder above the high tide line. "Let's sit there."

After Connor explained what he knew about the incident, I briefly gave my version and described my conversation with Betty Buttz.

"Angelo told me 'bout you wantin' to look into what happened."

I'd guessed that. Angelo and Connor talked about nearly everything. "I'd really like to hear Ryan's side," I said.

Connor shook his head. "Seen him a few times. Lord, he looks rotten. Too much drink, not 'nough sleep."

"The guilt must be terrible."

"'Course it is. What a thing."

"Do you think he'll talk to me?"

Connor ran stubby fingers through his curls, and I noticed that gray was beginning to win the battle against black. "What can I say? He's tryin' to work. You'll probably find him on the MOI dock."

Connor slid off the rock and squinted. "But you know, the lawyers. Ryan's not talkin'. Like Betty said, they want blame elsewheres."

We picked our way down to his boat in silence. The rising tide made the launch easy. After Connor jumped in, I shoved him off and yelled, "Hey, thanks for your help."

He revved the motor and was away, waving as he sped off.

As I trudged back to the Neap Tide for my own brew, I felt unsettled. Connor and Ryan were members of an unofficial Irish club in which loyalty was all. Connor might try to protect his friend. He'd quoted an Irish saying a while back. "Don't underestimate the undying love of one drunken Irish slob for another."

At the time, the quip was hilarious. Now, it wasn't so funny.

I got the coffee and spent the rest of the morning at my desk. Gordy was due at ten to talk about our next steps, and I wanted to reread the latest papers on New England fisheries and climate change.

At nine fifty, Gordy knocked on the door and stuck his head in. "Is it too early?"

"Come on in, Gordy, grab a chair." As he got settled, I asked, "Did you meet with the League fishermen, ah, folks?"

"Yeah. Tons of questions. Most 'bout this climate change business. They want the proof."

"Here's what I'm thinking. Start with what they know. Lobster distribution has changed in the Gulf of Maine, right? I'd like your pals to talk about what they've experienced. Then I can show water temperature data that supports what they're seeing."

"Gulf of Maine's a huge area. We're gettin' fewer 'n fewer lobster south, more up north."

"Right. From what I've read, lobster populations are shrinking down near Cape Cod and expanding up north. It's ideal for our needs. I can get fisheries data now and from decades ago."

"Not fish scientists' numbers, Doc. Like I said, fishermen trust those like they trust the gov'ment."

We decided I'd meet with a small group of League members in a few days to talk about the project—without fishery biologists' data. Our goal was to recruit eight or so fishermen to model how people who work the ocean and scientists who study it could effectively interact. With more specifics, my next job would be to locate interested scientists.

Gordy lifted off his baseball cap, smoothed his hair, and yanked the cap back down over his eyes. "That's it then? 'Til next week?"

"Actually, there's one more thing I'd like to ask you."

"Sure."

"It's about Ryan. I'd like to speak with him."

Gordy leaned back in his chair and crossed his arms. "No offence, but Ryan's not talkin'."

"But Peter's family. And me. We want to understand what happened. Ryan can help with that."

"Ryan's the best of the best. He did nothin' wrong. Lemme put it this way. If it was me, I wouldn't talk to my bosom friend. Rumors fly out like butterflies, come back like wasps, you know."

After Gordy left, I leaned back and looked up at the ceiling. Both Connor and Gordy said the same thing—Ryan wasn't talking. It wasn't surprising that Ryan was mum, but Connor and Gordy used nearly identical words to say so.

I organized notes, questions, and next steps for the grant proposal, then took the stairs down to Harvey's office. I needed her advice about a sticky problem. Harvey's door was open, but she wasn't there. I walked across the hall to her lab.

Hands on hips, Harvey stood before a contraption that occupied ten feet of lab bench. Comprised of six rectangular boxes connected by dozens of long, clear tubes, the auto-analyzer was a love-it or hate-it device. When it worked, countless little bubbles slid through the tubes, and between each pair of bubbles was a separate bit of liquid heading for analysis. On those days, Harvey could perform chemical measurements on hundreds of water samples.

Judging by Harvey's posture and mussed hair, I guessed it was an auto-analyzer hate-it day.

"Busy morning?"

Harvey walked behind her instrument, leaned on the table, and peered at something. "Damn thing's acting up again. Carol's on the phone with the company. We've got a pile of samples waiting in the freezer." Without looking at me she said, "What's up?"

"Do you remember that student, the one I encouraged last year when grant money looked better? Alise?"

Harvey fingered a piece of tubing. "Hmm."

I paced in front of the machine, gesturing as I went. "I was counting on that NOAA grant and can squeeze out summer support. But what about after that? Even if we get this NSF money, it'll be a while. In the next few days, I've got to figure how to pay her or tell her not to come. She'll be devastated."

I glanced toward Harvey. It looked like she was trying to trace the tubing as it disappeared into the bowels of one part of the instrument and reappeared elsewhere.

I continued pacing. "And the other thing is Sarah's email. There's no names on it, so how can I contact that sci-fraud group?"

I stopped and stuck my head behind the auto-analyzer. Squatting, Harvey frowned at what looked like a pump.

"Did you hear me at all?"

Harvey stood. "Sometimes the most obvious thing is right in front of your nose."

"What?"

She gestured toward the errant machine. "I see the problem. And yes, Mara, I heard you. But I'm too busy to talk."

"Jeez, Harvey, I'm sorry. I'm so, you know—"

"Preoccupied?"

I ran a hand down my ponytail. "I prefer to call it pensive, but you're probably right. My bad." I patted the auto-analyzer. "And good luck with this."

I reached the lab door. Harvey called out, "Try the organizer of the scientific fraud workshop at the meeting last summer. And talk to me later about Alise."

Back in my office, I fell into my desk chair. My arms felt itchy. No idea what that was about. I wore my usual cotton turtleneck under a fleece pullover, so it wasn't wool or something else scratchy against my skin.

I rubbed my arms and rolled my shoulders. Work I'd put aside was piling up. I had promised to review other scientists' grant proposals and papers. Equally pressing was analysis of the buoy data. Besides that, I had to prepare for the Fishermen's League meeting with Gordy.

Responsibility to the new graduate student, Alise, also weighed on me. My lab felt empty since my former students left for jobs, and I missed the collaboration. With everything that'd happened over the last week, I'd put financial support for Alise aside.

Then there was Peter. I got up and cleaned my whiteboard. Once more, I jotted possible leads and questions along with actions I could take.

Ryan's role? Talk to him.

Link between Peter's email and Frank Lamark?

Contact sci-fraud workshop organizer

I stood back and tapped the marker on my thigh. What else? One more name came to mind.

Cyril White? Visit him, see photos and video.

Frowning, I scanned the list. Pretty lean. There had to be more. What else had bugged me over the last few days? I added another name.

Georgina Hamilton Look at her profile again on Internet. Follow up?

Where to start? Cyril White was at the preserve, and I had asked to see his photos. There might be something interesting there.

I found the phone extension and called him. He picked right up.

"Cy, it's Mara Tusconi. Do you have any time today? Your photos—"

He interrupted me. "No, sorry. I'm super busy today."

"Tomorrow then?"

"Look. I'll contact you when I have time. Sorry." He ended the call.

I slowly put the receiver back. Cy was so friendly on the cruise, but for the second time in two days he barely had time for me. Maybe he had good reasons for his behavior, like a deadline or girl problems. I didn't have a clue.

At the whiteboard, I underlined the name at the top of the list. I'd deal with him after Peter's sci-fraud group. An hour and many phone calls later, I finally zeroed in on a scientist who "very likely" knew about the group. I called the likely candidate who wasn't available, naturally, and left a phone message.

Time for my next mission.

I spotted Ryan right away on the MOI pier. He was on *Intrepid*'s aft deck with his back to me. I walked up the gangway, my footsteps tentative and measured—nothing like the last time I boarded the ship, less than a week ago.

Preoccupied with what looked like a new overside crane, he didn't turn as I approached.

"Ryan."

He swiveled so fast he fell back against the railing. "Umph. Oh. Dr. Tusconi."

If you'd told me I was looking at Ryan's father, I would've believed it. Ryan's face was pasty, and wrinkles under his eyes drooped into flaccid cheeks. Hardest to look at, though, were his eyes. Pale blue, like pain had robbed them of their shine. And they searched mine as if I might offer relief.

"Sorry, didn't mean to scare you."

The old Ryan would have countered with "be-Jesus" in a Gaelic lilt. This Ryan only stared.

"Do you have a minute? To talk?"

He looked at the crane and coughed. "I'm done here."

"Well, how are you doing? I mean—"

He cut me off. "Like you'd think."

I bit my lip. "Ryan, I don't know what you're going through. But can I tell you why I'm here?"

He leaned back against the railing and crossed his arms. "Go ahead."

"You know Peter took my place. So I feel responsible, you know, to understand what happened. Does that make any sense?"

He stared down at his feet. "Suppose so."

"Can you tell me anything?"

He glanced up at me. "He told me not to talk."

"MOI's lawyer?"

"Yeah."

I waited. His tone was bitter, and I guessed he hated being told what to do by a lawyer with soft hands and fake nautical buttons on his suit jacket.

His speech was hard. "All I can say is that it *was* an accident."

"Ryan, I didn't mean—of course it was an accident. But why did it happen?"

"Said what I could."

11

THREE DAYS HAD PASSED SINCE I'd visited Angelo, so I dropped by on my way home. I found him in a sling chair on his slate patio overlooking the water. He stood as I approached.

"Hello, dear." He pointed to a companion seat. "Come sit."

I squeezed his outstretched hand and slipped into the chair.

In silence, we took in the predictably lively late afternoon sea off Spruce Harbor. With a fifteen-knot wind, white caps had replaced smooth-shouldered waves. Good thing I'd grabbed my fleece jacket from my car.

I spoke first. "I haven't seen you since my visit to the aquaculture place."

"You've certainly been on my mind. Was that worthwhile?"

"I'll tell you. But first, I just talked to Ryan. He looks awful."

"Being accountable for a death like that. My lord."

"I tried to explain why I wanted to speak with him. But he only said it was an accident. Nothing about why it happened."

"I'm not sure what you expected, Mara."

"Maybe something about the winch and inexperienced crew."

"Publically implicate MOI? He can't do that."

"I understand. But it's really frustrating. I'm not getting anywhere."

His smile was Mona Lisa-like.

"What?"

"You've many wonderful traits, Mara. But patience isn't one of them. You're like your mother that way."

Over the years, Angelo and I had shared stories about my parents, and I'd learned about two people I'd just begun to appreciate as an adult. Dad, the classic scientist, was cautious and logical. Mom, the adventuresome one, took the lead on environmental activism. When they died in the submarine, both had national reputations as marine conservationists.

"So what happened at the aquaculture place?"

I filled him in. As an engineer, Angelo was interested in the layout of the place and mechanics of growing algae in huge outdoor tanks.

"It was so frustrating the piers were off limits. But I saw them from above in Hamilton's office and can picture the set-up. I counted four cylindrical tanks on each pier, two side-by-side. They looked tall, maybe eight feet, and wider. Tubes the width of your finger ran up the side."

"Was seawater spilling over the top?"

I looked away and replayed the image in my mind. "A little."

"I've seen systems like that for growing fish. Those tubes up the sides must carry seawater from the bay into the top of the tanks at the same rate as water at the tank bottom flows out. It's hard to balance the two."

"Where does the bottom water go?"

"Very likely, it'd drip out a tube running through a hole in the pier."

"And down into the seawater?"

"Yes. Why?"

"Oh, just wondering."

I sped home, wolfed a sandwich, changed my clothes, and was on the road again in less than an hour, headed to the April swimming pool safety sessions at the YMCA with my sea kayaking buddies. The Y is west of town, near the high school.

In the parking lot Kevin—my pool pal—greeted me with a hug. His hairstyle had gone from preppy to spikey and it looked cute. He helped me lift my kayak off my car and carry it into the pool area.

"How was your winter, lovely Mara?"

Why Kevin calls me "lovely Mara" or his other version, "Mara, my love," I have no idea. But since I return the favor with "Stud-muffin Kev," I figure we're about even. Besides, what woman in her thirties minds being called lovely by a good guy who's ten years younger?

I told him I'd already gone to sea. "And you?"

"I guided river and rainforest trips in Costa Rica. Gorgeous."

Kevin had a reputation as a naturalist guide because he knew a lot about ecology, geology, and native peoples. Influential, rich people sought him out. Kevin relished the chance to talk with them about declines in the world's biodiversity.

"It's one thing," he'd said, "to read about extinction of animals like the golden toad from Costa Rica's cloud forest. But talk to locals who saw the iconic toad disappear in a few years? That's totally different. If there are politicians and wealthy business folks on my trip, I make sure conversations like that happen."

The pool area was already crowded with kayaks and kayakers, everyone psyched to once again practice rescues and rolling. The waters we paddle in, especially in April and May, are very cold—ocean water just warming up after the winter. Hypothermia, when core body temperature drops, can literally happen in minutes if we dump from our kayaks and can't get out of the water. All sea kayakers know gruesome stories of athletes who become paralyzed, lose consciousness, and die thirty feet from the shore. So every year, serious paddlers practice three techniques—self-rescue, assisted rescue, and rolling their kayak.

At the shallower end of the pool, people worked on rescues unaided or in pairs. A couple I recognized was halfway through guided rescue. The husband was in the water and wife in her boat. He treaded water and gripped the stern of his boat, which was perpendicular to hers and within her arm's reach.

"That's it," she said. "Rest the bottom of your bow in front of my cockpit. After we flip your boat, you can push down hard on the stern, and I'll lift the bow." She grabbed the front of his upside down boat and elevated it. Water from his cockpit ran out into the pool. Next, they'd maneuver the upright boats side by side, and she'd reach over and steady his so he could climb back in.

"Want to work on rolling?" Kevin asked.

"Yeah. But I'm nervous. Couldn't get the hang of it at all last year. Jeez, that was frustrating."

"I think you mentioned you had a bad time with the cold once."

"Right. To get my scuba certification I had to do an ocean dive in March."

"What happened?"

"It was unbelievable. Even with a thick wetsuit, gloves, and headgear, I was hypothermic in minutes. My whole body shook. I hyperventilated. Not good underwater. I sucked air, and my tank went down fast."

"Mara, if you paddle in the spring, you've got to learn how to roll."

Paddlers who can "roll" their boats don't have to eject themselves from their seats if they tip over. With their knees as braces, they use their paddles to roll the kayak right back up. But the so-called Eskimo roll is a lot harder than it looks, especially with boats seventeen-plus feet long.

A half-hour later, I hadn't rolled my boat once.

Gasping, I popped up to the surface, paddle in hand. My boat floated upside down, and I grabbed on to it.

"God damn! I just can't get the feel of it."

Treading water, Kevin explained my problem once more. "Stop trying to use the paddle to push yourself off the bottom, 'cause in the water you're paddling in there is no bottom. Relax in your seat, reach up, and time your sweep of the paddle with the hip snap."

I floated on my back to catch my breath.

"Try guiding the paddle up to the surface again. Except this time, do it a little slower."

Kevin waited patiently while I redid my single braid, clambered back into my boat, pumped out water, and reattached my spray skirt. Once more I tipped my boat over, sat upside down in the seat underwater with paddle in hand, and tried to relax with yoga breathing. Kevin reached down for my paddle and positioned the blade correctly. He let go. Now it was up to me to skim the blade across the surface of the water like a water ski.

All at once, the separate steps came together. As the kayak rolled to its side, my body came along, head floating just below the surface. I pushed down a bit on the flat paddle blade, shifted my hips, and popped upright.

Kevin whooped, and I joined him.

"Try it again?" he asked.

"You bet!"

By ten, I was one exhausted kayaker. I'd rolled a half dozen more times and dumped once. Not bad for a beginner.

After I showered and changed into dry clothes, Kevin helped me carry the kayak back to my car.

"Lots of people hate being underwater like that. How is it for you?" he asked.

"It's funny. If I'm in control, it's awesome. I can hold my breath for a long time, and it's so quiet under there. My mind clicks into a calm state. You know, like yoga."

We secured my boat on the kayak cradles. Kevin gave me a high five.

The next morning, a blinking light on my office phone greeted me. The recorded message was from Cliff Peadmont, who I hoped was familiar with Peter's sci-fraud group.

I returned the call, and he picked up on the second ring. I explained what I was up to. "Dr. Peadmont, Peter's wife asked me to look into this cryptic email. It was the last thing they talked about before Peter left for the cruise."

"Call me Cliff. I didn't know Peter personally. Was it an accident?"

That struck me as an odd question. "Why do you ask?"

"I'm not part of this sci-fraud crew. My thing is climate change doubters. I track their tactics by reading blogs and online forums. Lately they seem real anxious the public will actually listen to scientists. Petroleum interests and hardcore conservatives are giving doubter groups more and more money so they can escalate their game."

"What do you mean?"

"Those guys knew exactly what they were doing when they hacked the British university server. They delivered select phrases written by famous global warming researchers to a greedy press. Worked like a charm."

I shivered to think what the "greedy press" could have done with my own words. "But why would doubters target us at MOI?"

"You folks have some of the most important marine data on warming. Continuous temperature measurements in the Gulf of Maine? Coupled with what's happening to lobster? Hell, lobster are American as apple pie. Those little guys migrate out of the U.S. and up into Canada. Now, *that's* a story."

Impressed as I was with Cliff's understanding of our

research, I needed information. "Can you give me the name of the sci-fraud online discussion coordinator?"

"Gloria Sethman."

I jotted down her information, thanked Cliff, and tapped the numbers. Gloria answered right away.

Once more, I explained the reason for my call and recited Peter's words verbatim.

"I remember the email," Gloria said. "But the next group call was after the cruise, so Peter never explained what he was talking about."

My stomach clenched. "You've no idea what he meant?"

"I didn't say that. Actually, I saw Peter at a meeting right after he'd stumbled onto that genetic engineering situation."

I closed my eyes and placed a hand on my chest. "Yes?"

"The thing is, he didn't say much."

"What did he say?"

"Um, it wasn't the usual biomedical research. Give me a minute."

I grabbed a pencil and doodled.

"Something about sustainability. That's all I remember."

"You've been a terrific help. If anything comes of this, I'll let you know."

I hung up and scribbled a few words and phrases— *genetic engineering, close to home, scientist hasn't published,* and the new one, *sustainability.*

Huh. Sunnyside was all about sustainable fuel production. Maybe Peter *was* suspicious about Frank's *Chloronella* claims after all.

What was it Harvey said from behind her auto-analyzer as I pestered her with my worries? I could hear her voice in my head. "Sometimes the most obvious thing is right in front of your nose."

There was a big problem though. Sunnyside was a private business. If Frank was covering up a hoax, I had no way to find out.

I slid my computer on top of my scribbles. Suspicious algae would have to wait. I had to prepare for the meeting with Gordy's buddies. With that behind me, I'd be able to focus on the bits and pieces of my investigation into Peter's death.

By late afternoon, I was exhausted but amazed. Cliff was right. In the Gulf of Maine, fisheries data showed a warming trend even more dramatic than I'd expected. Half the thirty-odd stocks studied, including commercial species, had shifted northward.

And that included Maine's iconic crustacean. Lobster from the south joined their Maine cousins, creating a glut of lobster in the market. But as warming continued, lobster were bound to keep moving north into Canadian waters.

The Gulf of Maine was a poster child for climate change. If the folks who fished Maine waters refused to believe this, they'd die off with the dwindling lobster. On the other hand, it was human nature to deny what seems impossible. Gordy and I had to figure out a way to deal with that.

I shut down the computer, fell back into my chair, and fiddled again with hair I'd fiddled with all day. Time for a break. A hike up Spruce Mountain to clear my head was just the thing. It's not much of a mountain, but a steep drop-off at the top offers a pretty view of Spruce Harbor.

At the trailhead, I parked along the road and started up the dirt path. The first part's easy, so I pondered without worrying about tripping.

I tried to piece together disparate parts of my investigation into Peter's death. Just like with scientific data, a pattern might emerge. Fifteen minutes later, I was closer to the top of Spruce Mountain but not to the clarity I hoped for. I called Harvey.

"It's Mara. I'm on Spruce Mountain. It's lovely up here. Want to join me?"

"Lord, yes. My brain is fried. Be there soon as I can."

As the trail steepened, I watched for loose rocks. I was lucky to have Harvey. Besides seeing each other at work, we often shared weekend evenings. We'd both suffered through bad relationships and now lived on our own. When I first started at MOI, we barely knew each other well enough to say hello in the hallway. But the day I discovered my live-in boyfriend's last—and for me final—affair, Harvey found me sobbing in a bathroom. Breakups are awful for everyone, but my parents' deaths made me especially vulnerable to personal loss. Harvey waited until I stopped crying. We walked down the back stairs and across the gravel lot to the end of MOI's pier where we sat side-by-side, legs dangling over the water. It was August. We talked until dusk, and the full moon rose huge and orange across the bay. We'd remained soul mates ever since.

I waited on the granite outcrop at the end of the trail.

"Hey."

Harvey strode up the trail and hoisted herself on to the boulder beside me.

I squeezed her arm. "Sorry 'bout yesterday."

"Don't be silly. I was as preoccupied as you were."

"How's the machine?"

"It's working well—we're nearly caught up. You looked far away just now."

"Just thinking about my father. Have I told you about the first time he took me down to the running seawater tanks in our building?"

"Don't think so."

"I was five. You can imagine how enormous that base-ment room must've seemed. And loud. All that gushing seawater. My dad had to shout over the roar."

"What'd he say?"

"He pointed to little animals zipping around a tank and asked what they looked like. I put my arms out front, bent over, and said 'Superman!' You know how baby lobster hold their claws in front like that."

"I bet he thought that was hysterical."

"Probably. I asked what they ate, and he explained there were millions of teeny creatures we couldn't see."

"Mara the budding ecologist. Maybe that's why you're drawn to microscopic organisms."

"Could be."

We were quiet for a minute. Harvey said, "Anything new with the investigation?"

"I talked to Gordy and Connor. Both acted protective. Like Ryan had something to hide, and they knew it."

"Maybe they're just good friends."

"Could be. Anyway, Ryan looks awful. He's probably drinking a lot. He insisted what happened with the buoy was an accident."

"Suppose you'd expect that. And Peter's email?"

"Progress, finally. After a bunch of calls and emails, I reached the woman—Gloria Sethman—who runs that sci-fraud group. She remembered a snippet of conversation with Peter at some meeting."

Harvey swiveled toward me, her gray eyes wide. "And?"

"The bioengineering fraud wasn't biomedical. She said it had something to do with sustainability."

"Really? So maybe it's this Frank character after all?"

"I can't think what else it could be."

Harvey rubbed her perfectly turned-up nose like she does when she's thinking. "But still, what can you do about it? You can't even go out on those piers, never mind get some *Chloronella* to analyze."

"So frustrating. You know, close but so far."

I invited Harvey for dinner—it was Friday night—but she said she had other plans and started back down the path.

"Anyone I know?" I called out.

But she just kept walking. Again, I wondered if Harvey and Ted were seeing each other. But if they were, it was odd that she hadn't told me.

I looked toward the south end of the harbor at MOI. My parents worked there and loved it as much as I did. Maybe more. That Frank might lie about important research on marine algae would outrage them.

My parents. If they were in my shoes, what would they do?

I closed my eyes and envisioned my mother on the rock beside me. I remembered the timbre of her voice when she talked about scientific integrity the summer before I started college. "Never try to prove anything. Be observant and see what nature tells you. As scientists, we must be utterly honest."

I did suspect Frank was dishonest. But with no evidence, I was stuck. Even if I somehow got *Chloronella* without trespassing, the algae wasn't mine to analyze.

I headed down the path and stopped mid-stride. I looked back at the outcrop, then strode quickly on. My cell rang as I unlocked my car. Harvey. She was excited.

"I stopped by my office to pick up some new isotope data to bring home. My grad student's data."

"Wha—"

"You remember. He's collecting water from bays along the coast to determine if there's run-off from lawns and agriculture. Mara, he has data from Winslow Bay."

I held the phone tighter. "Does it show anything?"

"The isotope signature indicates a fertilizer source."

"But there are expensive houses around that bay. Run-off from fertilized lawns must go into the ocean."

"Don't think so. The signature is stronger *away* from shore, not like it's from a terrestrial source. Sunnyside's piers are the only thing in the bay. You'll want to verify this yourself, but it looks like the fertilizer is coming from those piers."

"That means—"

"Your science fairy god-mom came through. You don't need *Chloronella* samples to investigate Frank. You can just collect water."

12

Harvey ended the call, saying she'd see me in the morning. I sat in my parked car and drummed my fingers on the steering wheel. Spring peepers trilled in a nearby wetland.

Harvey's student's isotope data opened a new door on the Frank-*Chloronella* question. To verify if Frank added commercial fertilizer to the emerald green tanks, *I* could collect my own water samples along a line from shore to the Sunnyside piers and send them off for isotopic analysis. This wasn't unethical since seawater belongs to everyone.

It'd been nearly a week since Peter died, and my progress was pathetic. But if I discovered a hoax at Sunnyside, at least I could tell Sarah that this was the sci-fraud investigation Peter most likely discovered right before he died. That might be a bit of closure for her. I sure hoped so.

But I had to sample the bay without Frank seeing me. Ten minutes later, I'd worked out a plan. It was risky, and I needed input from a skeptic.

I walked into Angelo's living room and took in the warmth and smell of burning oak logs. Connor and Angelo sat in leather armchairs by the fire, arguing whether the Deceiver or Surf Candy was the best fishing fly for striped bass.

Angelo stood. I hugged him, lingering in the softness of his favorite flannel shirt. Connor carried a chair to the fireplace for me. I patted his back in thanks.

"Hope it's okay I dropped in."

"Anytime, Mara, you know that," Angelo said. "We've already eaten, but lasagna's still out if you want to warm it up."

"Perfect. Back in a bit."

When I returned, the pair put aside the Deceiver/Surf Candy question. I slipped into the empty chair and the three of us chatted about new commercial fishing regulations.

Connor said, "I know those fish biologists are tryin' to do right. But the fishermen can't keep up. First this, then that. There's gotta be a better way."

Since we'd discussed the knotty problem before, I changed the subject. "It's been a week since Peter died. There's an idea I'd like to pass by you."

Angelo crossed his legs. "Of course, dear. Go ahead."

I described Peter's sci-fraud email, why Sarah asked me to look into it, and why Harvey and I doubted the super-seaweed claim and Frank's ability to genetically alter *Chloronella*. I glanced at Angelo but couldn't read his poker face.

"Wait," Connor said. "Why is this Cholor, whatever it is, so ungodly green if Frank didn't mess with its genes?"

"That tripped us up for a while. We're guessing Frank *thought* he inserted the nif gene because the algae grew well at first. But something happened, and *Chloronella* wasn't super after all. He panicked and kept people away from the piers unless he knew them. By then, he'd attracted investors with his claim and needed more money to keep going. We think he's adding commercial fertilizer to the tanks to keep his investors happy in the hope he can figure out what's going on."

Angelo asked, "But you don't think he succeeded in the first place?"

"We're not sure. But Frank's claim might be what got Peter's attention. Biologists have been trying to genetically engineer algae to fix nitrogen for a long, long time. Lots of agro-engineering labs. Peter may well have thought it was unlikely that an unknown biologist succeeded."

Connor's eyes widened. "Are you thinkin' this algae business and Peter's death are connected?"

"That's a real stretch. If there's a connection, I don't know what it would be."

Angelo held up a hand. "Hold up. I can't sort this all out. You're looking into this email because Sarah asked you to? Is that right?"

I leaned forward and held his gaze. "Something fraudulent and local incensed Peter, and he hardly ever got angry. Peter and Sarah talked about it right before he left. Trying to find out—it's the least I can do for her."

Angelo leaned back and crossed his arms. "There's more to this. You said you had an idea you wanted to talk about."

"Coming to that. I can't test *Chloronella* directly, so I've been stuck. But Harvey has new information that changes everything. One of her students samples bays along the coast to see if nitrogen from fertilizer runs into the ocean. Turns out, Winslow Bay, where Sunnyside is, shows a strong isotope signal. What's unusual is that the nitrogen source isn't terrestrial. So it might be Sunnyside's piers."

"I need some help here," Connor said. "What's an isotope? Is it radioactive?"

I turned to face him. "No, that's why they're called stable isotopes. Elements like oxygen and nitrogen—in algae, people, fish, whatever—come in two or more forms."

"They're the isotopes?"

"Yeah. Same element with different numbers of neutrons in the nucleus."

Connor nodded, so I kept going.

"If Frank adds commercial fertilizer to the *Chloronella* tanks, the nitrogen isotope value in water samples nearest the pier will point to that as the nitrogen source."

Excited, Connor stood up. "That'd mean the whole thing is a hoax. Sunnyside wouldn't be able to claim they've created a super-seaweed, or make big bucks growing it for biomass. 'Course it's bizarre. Why in hell they'd set up this fraud in the first place."

I shrugged. "Million dollar question."

"Sorry to interrupt the chemistry lesson," Angelo said. "But how can you get these samples without Frank finding out?"

I stood. "I have to collect water at evenly spaced points along a straight line from shore to the piers. Not much, just one hundred fifty milliliters at each point. I can kayak to the piers and sample along the way."

"But someone will see you," Connor said.

"Not if I paddle out in the dark."

Angelo looked up at the ceiling, then at me, eyebrows like question marks. "Mara, a solo kayak paddle out to those piers would be risky. For one thing, the water's still in the high forties. If you go over—"

"I've never dunked."

"Drysuit?"

"I'll be wearing a drysuit."

"You can roll the kayak reliably?"

I stretched the truth. "Yes."

He grabbed a poker and stabbed at the fire. Sparks sailed up the chimney. "Of course, I understand your sympathy for Peter's wife. But alone at night in a kayak this time of year? It's not worth the risk."

"It's not just getting samples. When I visited, the place felt...I can't explain it. My intuition tells me something's off at Sunnyside. Maybe I'll see something on the piers. I don't know."

Connor hit his knee and guffawed. "Intuition? Come on, you're a scientist."

The poker clanged against its stand. Angelo said, "I'm getting some coffee. Either of you want any?"

Connor and I declined. As Angelo left the room, I called out, "Can I help?"

I interpreted his wave as a negative.

"Give him a little time to chew over this kayak caper," Connor said. "While we're waitin', tell me 'bout intuition."

I relaxed into my chair. Connor was a great guy. While he might be interested in my explanation, I guessed he really wanted to ease the tension in the room.

"Most people think scientists are logical thinkers who pooh-pooh intuition and hunches. But I know of a few famous ones who paid close attention to a sixth sense, gut feeling, fleeting images. Whatever you call it."

Connor stroked his five o'clock stubble. "Give me an example."

"The classic's James Watson. He dreamed about two intertwined snakes before he and Francis Crick came up with the double helix structure for DNA."

"Amazing," Connor said. "It's like the unconscious makes pictures to give the thinking part a hint."

Angelo reclaimed his chair, a mug of coffee in his hand. He took a sip. I waited.

He spoke, his words measured. "Couple of things. Whatever you sensed as off, I think that's your word, might be dangerous."

"I'll paddle out to the pier and back. Nobody will see me."

He held up his hand. "The water's extremely cold. Even in a drysuit, if you go over, that water will hit you like a ton of bricks. A shock you can't imagine. I'm a marine engineer—I know."

"But I've never tipped over."

"You've got to understand what you're getting yourself into, Mara. Put the drysuit on and immerse yourself in the ocean."

If Angelo assumed this would quell my enthusiasm, he was wrong. "I can do that."

Connor clapped his hands together. "My god, Mara. You've got the spunk of the Irish."

Angelo glanced sideways at his buddy. "I can't tell you what to do and won't try. But give it another day before you make your decision."

His tone—deadly serious—chilled me. "I'll certainly do that."

The three of us filed into the kitchen, a bright, warm room Angelo designed. Four cabinets across one wall face two on both sides of the gas stove. The soft yellow cabinets contrast nicely with the black of the stove.

My godfather likes order and balance.

I pecked Angelo on the cheek and said good night to Connor. As I closed the kitchen door, Connor claimed to Angelo that Surf Candy was the gold standard for saltwater flies.

I pulled into my driveway, the night fully dark. Reminding myself again to fix the motion-detector spotlight, I picked my way up the stairs and across the deck to an unlocked kitchen door. I'd lived alone for years and relished moonless evenings like this because the starscape from the beach is astounding. This night, however, I felt edgy until I was inside my brightly lit kitchen.

Angelo's negative reaction had unsettled me. I should have expected his response, but I adored my godfather and we rarely argued.

I'd caused him angst and that felt lousy.

The next morning, I opened Spruce Harbor Town Hall's front door with one hand and held my computer projector with the other. Clutching the projector to my chest, I tried not to trip on the worn wooden stairs leading down to the basement room Gordy had picked for the Fishermen's League meeting.

Gordy greeted me from underneath a table. "Mornin'. Gimme a minute to plug in this extension cord."

I fiddled with the projector and attached the electric cord Gordy handed me. The Town Hall had no internet connection, so I had to rely on ancient technology. Gordy scraped two folding chairs across the floor. Knee to knee, we went through our agenda.

"First thing, Gordy. I'd rather you didn't call me Doc in front of your colleagues."

"Trade ya. Don't call 'em colleagues, and I won't call you Doc."

My chuckle felt good. I was more nervous about the upcoming give and take than I wanted to admit.

Right on time, the League members—seven men and one woman—ambled in and pulled out chairs around the U-shaped table. We set up that arrangement since I didn't want to be seen as a teacher who lectured from the front of a room.

I expected fit twenty- to thirty-year-olds. But apart from baseball caps, the group appeared to have little in common. There was a surprising range of facial hair—muttonchops, full beards, a walrus mustache—and the oldest looked seventy; the youngest likely couldn't legally drink.

Gordy explained the end game—a proposal to the National Science Foundation for lobstermen and scientists to anticipate the future of Maine's lobster fishery. I hadn't expected enthusiasm, but a little interest would do. I guessed the prospect of a salary was the carrot here.

"Dr. Tusconi, Mara, thought you might have questions about global warming in the Gulf of Maine. So—"

A man with a red square-cut beard interrupted. "Bes' lobsterin' used ta be down Portlan'. Now it's way up the coast." He slapped the table. "But who knows what's causin' it? And if it's warmin', is it gonna reverse? That's the question. If it's sunspots, they go away."

Heads nodded around the table.

"Sure," I said. "Great question. NASA satellites have measured solar output since the late seventies. They show that solar activity has actually *decreased* at the same time atmospheric temperature has increased."

Over the next hour, the group peppered me with questions.

"It's crazy to think people could change the climate."

"Antarctica's getting cooler."

"Scientists disagree about global warming. It's all a conspiracy."

"Last winter was the coldest ever."

With each response, I scanned the group to gauge the reaction. Arms crossed over chests was the norm.

They did chortle when I explained the difference between climate and weather. "Imagine walking a dog. It weaves all over while you go in a straight line. The dog's like the weather. All over the place, but it gets there eventually. And you're climate. Slow and steady in the same direction. We're talking about climate here."

We took a quick break at ten. I wanted a read on the group's mood from Gordy, but he was in the middle of a clutch talking about a union for Maine's lobstermen.

After the break, I asked if anyone knew about lobstering in other parts of New England.

"Long Island, it was like someone flipped a switch. Hot water 'n that shell disease. That was it for lobster and lobstermen."

"And down Mass'chusetts, it's an underwater desert."

"So," I said. "If it happened there, why won't it happen up here?"

They looked at me like I was crazy. "'Cause we got clean, cold water up here. Down there, who knows what them lobsters were livin' in."

The woman said, "You stupid or somethin'?"

That remark brought Gordy to his feet. "Hold on. Dr. Tusconi's only askin' questions. She's jus' tryin' to help us think about this stuff. No need to get nasty."

Before we broke up, I had one last question. "What do you think about working with scientists who study warming in the Gulf of Maine?"

Dead silence. Before, you couldn't shut these people up. Now they looked at their hands, out the window, at the ceiling, anywhere but me. I asked Timothy, the guy with the red beard.

"Our 'sperience with fish scientists is piss poor. They think we're dummies." Maybe I looked deflated, because he added, "I have ta' say you seem diff'rent."

At noon, Gordy and I were alone again in the silent room. Two people—the woman and Timothy—briefly stopped to speak with me on their way out. But the others simply left, chattering in twos and threes.

I thanked Gordy for stepping in. He waved his hand and said, "You're my favorite Irish aunt's daughter."

"It didn't go so well, did it?"

"Load of bullocks."

"But—"

"Hey, these are busy guys. Five minutes into it, they would've walked out if they didn't see worth in this thing."

"But Timothy, he practically said he hates scientists. If they don't want to work with us, the whole thing's off."

"Nah. You're diff'rent. That's what he said. That's a huge compliment, Mara."

Harvey's office door was open. Many scientists worked over the weekend, so that wasn't surprising.

I stepped inside. "Got a minute?"

It looked like she was reorganizing her bookcase. "What's up?"

"I've got an idea."

"Sounds like I need to sit down for this."

I described the kayak plan. "If I leave at dusk, I can paddle the mile or so to the docks and take samples along the way. It will be dark by the time I get out there."

"And cold!" Harvey protested. "It's only April. The water's winter temperature. You're always talking about hypothermia and danger of paddling alone. You could die if you fell in."

"You're right, but it's a short paddle. I'll wear a drysuit and pick a calm night."

"Does Angelo know about this?"

"He did his best to talk me out of it. And said I should immerse myself in the ocean to feel what that's like this time of year."

"You'll do it?"

I shrugged. "Sure."

She flicked a pink-polished nail and frowned. "Mara, read up on negative traits of creative, smart people. Dogged's gotten more than a few into a whole lot of trouble."

13

Harvey walked over to the window and stood, looking out. She ran her fingers through her hair a couple of times.

She sighed. "I told you about the isotope data just yesterday afternoon. Already, you've got a plan."

"Time's flying by. I'm still uncertain about Peter's email, never mind his death."

"This kayak trip—you can't do it alone."

"You're not an expert paddler. Who else can I trust?"

"I'll go and help with the drysuit or whatever. It'll be better than sitting home worrying."

The lump in my throat surprised me. "Would you? That'd be terrific."

"When do you want to go?"

"Maybe even tonight. I'll check the weather and call you."

Back in my office, someone knocked on the door as I stood on a chair and slid the computer projector across the top shelf of my bookcase.

"Door's open."

Ted stuck his head in as I stepped off the chair.

"Ted. How's it goin'?"

He gestured toward the window. "Hey, it's nice weather. Want to sit outside for a bit?"

After my morning with cranky fishermen, that sounded good. I joined Ted at the end of the MOI pier. He leaned against a piling, his legs outstretched. I took off my shoes and dangled my legs over the edge.

I said, "I remember sitting here with my dad when I was little. He'd throw a bucket down, haul it up, and carry it back to his lab. We looked at the water under a dissecting microscope. The invisible world, it came alive. Copepods with their big antennae and jerky movements, strings of algae floating around like silky threads."

"What a nice memory." Ted paused. "I know your parents helped lead the way on marine conservation. But I don't know what happened to them."

I swung my legs up on to the pier and blew out a long breath. "They were among the early biologists in the fifties and sixties who used subs to study the ocean. Shallow dives three hundred feet down on coral reefs. During a dive off St. John, a hatch seal failed at the same time the sub got wedged between two coral mounds."

I glanced at Ted. His cobalt eyes were fixed on me.

"Backing out from the mounds would normally have been routine." A clump of seaweed rocked back and forth with the current. "But seawater poured in and my parents and the pilot drowned before they could do it."

"Goddamn," he said. "How old were you?"

"Nineteen."

We both stared at the harbor. He said, "We never think much about it, but what we do for a living can be dangerous."

I opened my mouth to say something about nighttime kayaking in April but shut it and bit my lip.

I stood. "Got to go." Waving at the harbor I said, "This was a really nice idea."

Driving home, I pondered why I didn't tell Ted about the Sunnyside kayak idea. Or that Frank might add fertilizer to the tanks. After all, Ted had revealed confidential information about Frank and the nif gene.

I didn't confide in Ted because he would protect his friend. There might be other reasons that hadn't worked their way through my subconscious.

I carried a bag full of gear down the footpath in front of my house to the beach. Time to make good on my promise to Angelo. I plopped the bag onto wet gravel near the water's edge and spread out a beach towel. Then I began the task of pulling the drysuit over my clothes. For anyone watching, my gyrations and contortions would've been the day's entertainment.

I waded into shallow water and felt nothing; the drysuit insulated me from the cold. The tide was low, and I had to go out a ways before the water got deep. I took a big breath, jumped up, and slid down beneath the surface.

I came up fast with a yelp.

The jolt of pain on my face astounded me. My head felt like it was in a vice grip made of ice. I gasped, staggered backward to get out of the water, and plopped down on the rocky beach, panting. My face burned for a good five minutes.

I gathered up my things. Angelo was right. I hadn't been prepared for the shock of April seawater in Maine. It was no picnic, but now I knew.

In the shower, I turned the water to full hot and stood there, head under the faucet. Then I rubbed my arms and legs hard with my scratchiest towel and pulled on two layers of fleece.

I checked the weather forecast for that evening. I liked what NOAA told me.

I called Harvey. "The winds tonight will be calm and air temperature above normal. Tomorrow's Sunday, so we can sleep in."

"I'm all yours."

It took a long time to assemble the cold-weather kayak gear on my deck. The drysuit, my thickest hood, insulated paddle gloves, booties, PFD—Personal Flotation Device— with new reflective safety tape, spray skirt, compass for the boat, plus a fleece hat. I could wear fleece long johns under the drysuit but needed a change of clothes in case the fleece got wet. Those went on top of the growing pile.

More than once, Angelo had teased me about bobbing around on the ocean in a long, narrow, tippy boat. Looking at all the stuff cold-water paddlers need, it was easy to sympathize with his point of view.

The kayak racks were already on my car from the pool session. I carried my boat from the garage, slid it up onto the racks, and tied it down. At seventeen feet, the kayak is awkward to move around, but I can lift its fifty pounds with no trouble.

Usually, my heart sings in spring when my kayak finally slides into the ocean. But today would be entirely different. I had no idea what kind of a first trip it would be.

In mid-April, the sun in Down East Maine sets around seven. I'd calculated the time we needed to drive north to Sunnyside, unload the kayak and gear, and get me into the drysuit. Harvey said she'd be ready at five-fifteen.

Dinner was a ham sandwich—some New Englanders say pork's supposed to be good luck. I drove my packed station wagon into town to pick up Harvey. Next to her garage were two canvas tote bags filled with stuff.

I pulled up her driveway and stepped out of the car. "You're not going away for a week."

"Who knows how long I'll wait for you." Pointing to each bag in turn, she said, "I've got something to eat and drink, papers to review, my reading group's book, and warm clothes. Even hot chocolate for when you get back."

On our way out of Harvey's neighborhood, we waved at a fellow professor walking her dog. Along Water Street we passed the Neap Tide, where a cluster of the gray-haired set gathered outside for meatloaf night. The table overflowing with used books was still in front of the bookstore, so they hadn't closed yet.

It all looked completely normal. The contrast between ordinary-looking Spruce Harbor and my far-from-ordinary mood felt strange.

As we turned north onto Route 1 Harvey asked, "How will you find your way to the piers as it gets dark? And what about the samples? You need to know exactly where you're collecting them."

"I plotted the route on my NOAA chart—you know, to follow with the compass. Each time I take a sample, I'll enter the waypoint on my handheld GPS. I've gone through each step of the trip in my mind a dozen times."

"Being there in person in the fading light will be very different."

I swallowed hard. "You got that right."

An hour later, I pulled off Route 1 and drove east a few miles to the shore road where we headed north again. As we neared the Sunnyside facility, I slowed and soon spotted an overgrown dirt road between two old maples. I pulled on to it. We bumped along for a few minutes and stopped with the ocean just in view.

Harvey looked at the beach, then over at me. "How do you even know about this spot?"

I shrugged. "I'm always looking for someplace to get my kayak down to the water without carrying it too far. Winslow Bay's protected from north winds. I paddled out of here last fall."

I slid the boat off the back of the car into Harvey's hands and together we carried it the short distance down to water's edge. It took several trips between the car and the kayak to get all the gear lined up on the beach.

"Boy, you have a lot of stuff here," Harvey said.

"And I need every bit."

I put down a beach towel, stepped onto it, and began the business of pulling the drysuit over my long underwear. First, I sprinkled baby powder on my feet, pulled on bootie socks, and sat down. The drysuit's rubber ankle seals were a tight fit, and Harvey helped me tug them on. I stood and slithered the suit up over my butt, which took another few minutes. Harvey sprinkled more baby powder on my hands, which helped me slide them through the arms and wrist gaskets. I shimmied the suit up on to my shoulders.

I paused to catch my breath. So far so good, but the head gasket was by far the hardest part.

Facing Harvey I said, "Okay. Reach over and position the neck over my head, then help me pull the neck seal down. It's super tight so be careful."

Since I couldn't see Harvey, and she hadn't done the procedure before, progress was slow and painful. After numerous ouches and complaints about pulled hair, my head popped through. I pulled up the drysuit zipper to finish the job and sat down once more to tug on my booties. The gloves would go on last.

Kneeling alongside the kayak, I adjusted the compass so it would be easy to read. Directly in front of the cockpit, I firmly attached a waterproof bag holding twenty acid-cleaned numbered sampling bottles, a waterproof flashlight, and my handheld GPS unit. Standing, I stepped into the spray skirt and pulled the straps over my shoulders, swung on the PFD, and zipped it up. Harvey snapped the two halves of my kayak paddle together.

I was nearly ready. Harvey helped me zip a spare flashlight, cell phone, power bar, and waterproof watch into pockets on my PFD.

We went over the schedule one more time. "Normally, it would take less than forty-five minutes to reach the Sunnyside piers. There's no wind and little current tonight. But I've got to stop and get the samples along the way, so I don't know how long the trip will take."

The sun was setting, but visibility was still okay. I walked the kayak into shallow water and slid into the cockpit with my paddle as a brace behind the seat. It was a snug fit, and the drysuit made it even more so. I wiggled into the seat, reached forward with my legs, and positioned my feet into the braces. With the spray skirt tightened around the cockpit, I pulled on my gloves and secured the Velcro straps around my wrists.

I pushed off and leaned forward toward my goal, body tingling with anticipation. Finally, I was in my boat heading out to Sunnyside. Finally, I'd get close to the off-limit piers and forbidden tanks.

Of course, I didn't know if the water samples would help Sarah understand Peter's confusing email. And we might be wrong about Frank and the fertilizer. I shook my hooded head. This wasn't the time to worry about that.

The lack of wind and waves made paddling easy, and I quickly assumed a rapid rhythm, alternating placement of the paddle on either side of the boat. One-two-three-four-five-six. One-two-three-four-five-six. Not far out, I stopped paddling to get the first sample. I unzipped the bag in front of me and pulled out a bottle labeled "1." I filled the opened bottle with water from just below the surface, screwed on the top, stashed it back in the bag, and pushed the waypoint button on the GPS. One down and nineteen to go.

In between my sampling points, the kayak slid onward, fast and sleek through the water. As I took sample number three, I looked toward shore. Harvey was a tiny figure on the beach.

I turned back. The compass was now barely readable in the fading twilight. A light fog settled around me, wet and cold, making navigation by sight impossible. I used my flashlight to read the compass bearing I'd worked out.

Soon, fog and darkness engulfed me. Just as the kayak's bow parted the soaked air, murk closed in right behind me. I was alone in a muffled-gray absolute—the swish, swish of my paddle my only companion.

At waypoint number twelve, unease set in. I should be close to the piers by now. What if my navigation was off? Paddling through endless fog to nothing was unthinkable.

When I slid bottle number thirteen beneath the surface of the water, I squinted through the fog. Still nothing.

At waypoint fifteen, I lifted the paddle out of the water. The kayak slowed, then slid to a stop. The fog was less dense now. I squeezed my eyes shut, opened them, and squinted once more.

Relief flowed through me. Directly ahead, faint pricks of light pierced the fog.

I stored the sample and paddled on with more energy. Quick, short strokes. Pin-pricks turned into soft glows that ran along the water. As the kayak slid to a stop at the next collecting point, ghostly forms of the docks and algae tanks emerged in the gloom.

I was nearly to the Sunnyside piers.

A couple of strokes later, a string of long white blobs bobbing in the water surprised me. Ah, the weird floodles.

At waypoint nineteen, I could clearly see a series of long piers jutting out from the lab building. Bright yellow lights exposed a man-made biology. In two rows running down the piers were the stout cylindrical algae tanks. From my vantage on the water, they were enormous, larger than they appeared to be from Hamilton's office window. I looked for signs warning intruders to keep away. There were none.

Although it was fully dark, various colors in the tanks were distinct beneath garish lights—dull tints of brown, red, yellow—at regular intervals down the docks. The emerald *Chloronella* tanks sat on the easternmost dock. Beneath the cold lighting, millions of algae cells inside each tank gave off an eerie green glow.

Gazing on it all, I felt foreign—a trespasser.

My last sampling point was just seaward of the pier. The kayak slid slowly to a stop below the tanks. I looked up and surveyed the creepy scene. What I guessed were ten-foot-tall and twenty-foot-wide neon green cylinders cast ash-gray shadows beneath yellow lighting. Bobbing in the water below them, I felt small and vulnerable. The only sound was a gurgle of water mixed with the drone of mechanical pumps.

The place felt sour, almost evil.

After all the planning and deliberation, I'd made it here—and now wanted to leave as quickly as possible. I felt exposed, like someone was watching me. I looked around. I couldn't see any surveillance cameras.

I took a deep breath, then blew it out.

Water for waypoint twenty bubbled into its bottle. I slipped the sample into my bag. There were four unmarked bottles left.

To verify how *Chloronella* water entered the bay, I needed to see tubing that exited from the bottom of each green tank. I slid the kayak underneath the pier and pulled the flashlight out of my PFD pocket. The tubes were exactly where Angelo guessed they'd be. Spaced about twenty-five feet apart, tubes the width of my thumb stuck down below the pier. Greenish water dripped out of each one into the seawater. I counted four of them.

Even though winds were calm, the water surface beneath the pier was what kayakers call squirly—the boat moved unpredictably in a current going this way and that. Maybe the sturdy pilings supporting the pier obstructed water flow. Whatever the reason, the current buffeted the kayak more than I would have liked and made control of the boat difficult.

I decided to get some seawater from each exit tube. As I screwed the cap on the third one, a distinctive sound startled me. I heard it clearly above the gentle slaps on the pier.

A door had opened and slammed shut at the lab end of the dock. Footfalls on the pier followed.

My hands shook as I dropped the bottle in the bag and zipped it up. I closed my eyes and forced myself to take slow, steady breaths.

The realization that I was safely hidden from anyone on the pier helped me calm down. Curiosity took the place of fear. I desperately wanted to know who was walking toward me. With the stern of the boat poking out from beneath the pier, I figured I could raise myself up just enough to see the deck and maybe a person. They wouldn't be looking for anyone down in the water, so I'd be safe.

The footsteps clunked down hard like a man's. They stopped.

The kayak didn't want to stay put and nearly banged into a piling. I strained to straighten the boat as it moved around. *Come on. Come on.*

Footsteps began once more as he walked closer and closer toward me. I paddled backward a bit and pushed myself up.

About thirty feet away, a man passed beneath one of the lights. He was blond, tall, medium build, chiseled.

My stomach did a flip. It was Ted!

I nearly cried out as the boat slid back beneath the decking. Holding the paddle in the water to steady the kayak, I trembled violently. It might've been fear, anger, or both.

Although Ted wasn't visible now, the pier shook as he walked up to each *Chloronella* tank. He was doing something to each one. Maybe adding fertilizer.

It was almost impossible to hold the boat steady in the squirly water, but I had to. If the kayak smacked against a piling, Ted would certainly hear it. He was right above me, after all.

As he stopped at each tank, there was a loud, dull clunk. It sounded like wood on wood, so I guessed he dropped a wooden box beside each tank and stepped up on to it. After the clunk, all was quiet. A few moments later, it sounded like he stepped off the box, picked it up, and moved on to the next tank. He passed directly over my head. I shucked in a breath even though the slapping of waves against the pier would muffle such faint sounds.

I counted clunks. At number four, Ted would be at the last tank. The box clunked down.

At that moment, a wave came out of nowhere and slammed the boat against a piling. I covered my mouth to stifle a scream.

14

I DIDN'T MOVE AND NEITHER did Ted. Dear Lord, make him think a piece of driftwood hit a piling.

God wasn't with me—a run of waves lifted the kayak and smashed it against two pilings. I was helpless to do anything.

Ted's curses—furious and harsh—chilled me to the core. Terror bolted through me. How could I explain what I was doing there? Even if I did, there was no way to know what Ted would do.

He stomped around, swearing under his breath. My mind ran through my options. There were only two. I could stay put or push off and paddle away.

An intense beam of light played across the water at the end of the pier. It looked like Ted had turned on one of those big emergency flashlights. Vertical slits of light slid over my kayak. I nearly whimpered.

Time to get out of there, pronto.

I poked the boat out from beneath the pier and used a piling to push myself in the direction of shore. Stroking as swiftly as possible, I paddled away.

I blessed the light fog and my headgear. Ted would notice the boat, but a kayaker fading into the murk would be all he'd see.

He spotted me instantly "Hey! Get back here! Son of a bitch!"

Stroking hard, I pulled away—ten feet, twenty feet, thirty feet. The sound of Ted's voice was muffled now. With a whispered, "Whew," I blew out the breath I'd been holding. I relaxed the muscles in my arms and flexed my fingers a few times—and tried to paddle briskly and efficiently away from Ted.

Moving along well, I congratulated myself on my speedy escape.

That's when I heard it. The distinct roar of a motor. How incredibly foolish I was. It never even occurred to me that there'd be a motorboat tied to the Sunnyside pier.

I tried not to panic but knew my life might be in danger. A kayak moves a few miles an hour. Ted probably had a big motor that could propel a lightweight boat twenty-five miles an hour. Within minutes, he could overtake me. Who knew what he'd do.

I kept paddling and changed direction, praying that he wouldn't be able to find me in the fog. The motor droned on, closer and closer. And like the intense beam of a light-house, the beam from his powerful flashlight swept back and forth, back and forth, cutting through the murky dark-ness. It probed in every direction—toward the shore, out to sea, back to the pier.

Suddenly the shaft stopped dead, and I was bathed in the spotlight. The beam had found me.

The whine of the motor grew louder. Ted had turned his boat to follow the beam.

Now he sped right toward me.

My mouth went dry. The reflective tape strips on my PFD, a safety precaution so other paddlers could see me in a fog, had turned into a hazard.

Ted quickly gained on me like a hungry shark. Given his speed, I felt certain he was going to ram my boat.

My little kayak would be nothing to a powerful motor-boat. He'd run right over me. The grinding blades of his motor would shred the kayak and rip into my body. As if in a movie, I saw it in slow motion. The terrible force of the collision, my boat shattering around me, my screams, the hot pain, gasping for air, the ultimate blackness.

Ted was very close now—less than a hundred feet away. There was no way out.

Emerging through the murk directly in front of me were the weird pods—the floodles—white oblong tubes all in a row, bobbing in the water.

Upside down, the bottom of my kayak would perfectly mimic a floodle.

I maneuvered my boat next to an end pod and positioned myself parallel to the row. The roar of the motor filled the dense air, and the flashlight probed the gloom.

I sucked in a deep breath and rolled the kayak.

The arctic seawater hit me hard, astonishingly cold on my face. Eyes shut tight, I was engulfed in the fire of ice.

Hanging upside down, I fought the intense urge to suck in air.

It's the reflex to sudden cold. You don't need to breathe.

Panic bubbled up.

Could I roll back up? Rolling the boat in these hellish conditions would be a whole lot different than in a pool.

I forced myself to retreat into that tranquil-beneath-the-water state of mind and focus on Kevin's calming voice in my head.

The sound of the boat motor was much, much louder now that I was underwater. As the drone intensified, my body vibrated with it. Kevin's voice fought for my brain's attention.

Be cool, lovely Mara. Picture your paddle setup.

The motor was so piercing I was certain Ted would run right over me. And with the adrenaline racing through my body, my lungs screamed for air. I *had* to suck in water.

In an instant—as quickly as it had escalated—the cruel drone faded. Ted flew by the pods and raced on.

With my last bit of energy, I looked up, positioned my paddle, and rolled right out of the water.

A perfect roll.

Once upright, I had to get the hell out of there—leave the motorboat behind and get to the safety of shore. My face stung horribly, my head throbbed, and my brain was wooly-sluggish. I desperately wanted to be back on the beach with Harvey but wasn't sure about the direction. Urgently needing to move, I struggled to turn the boat away from the sound of the motor.

My paddling was clumsy at first, a chore just to lift the paddle up and drag it back through the water. Right side—now left side—right again—left. Pause to rest. Right—left—right—left.

I stopped to catch my breath and listen. Ted's boat was only a distant hum in the fog. I pulled the flashlight out of a PFD pocket, turned it on, and bent over to read the compass mounted on the front deck. My brain kicked in. I adjusted my direction to the correct setting and paddled with steadier strokes.

Blessedly soon, I could hear pebbles rolling up and down the beach with the waves. I made out a dim light that moved back and forth. Harvey's voice called me.

The kayak slid up out of the water and onto the sand.

I was safe.

Harvey hauled the kayak higher up the beach. I tried to push myself out of the cockpit, but my trembling legs wouldn't support me. "Damn, I need your help with this."

Harvey had seen me exit my kayak with no trouble. She quickly straddled the boat behind the seat and linked her arms beneath my armpits. "Push with your feet while I pull up."

With her help, I stood and stepped out of the boat. Steady on solid ground—but dripping wet—I swung around and hugged my loving, dependable, generous friend.

A half-hour later, I sat in the passenger seat of the car, holding my hands against the heating vents. All the kayak gear—wet and sandy—was in the back of the station wagon.

The water samples were on ice in a cooler. Warming up, I sipped hot chocolate, tremendously happy to be heading home.

After a quick call to Angelo, who said he'd phone Connor, I described my adventure to Harvey. "He was a crazy man, incredibly angry. When he started swinging that flashlight beam around, it was time to get the hell out of there."

"That's *not* Ted. It doesn't make sense," Harvey said. "We're on a committee together. I've never seen the hint of an angry nutcase." She glanced over. "Mara, are you *sure* it was Ted?"

I closed my eyes and pictured the scene once more. "Yes. The lights were very bright. It was Ted, no doubt."

Harvey didn't argue more, but it did bug me that she questioned what I saw with my own eyes. It would make sense that she'd defend him if they were going out.

As we zipped along the road I described the motorboat chase, my quick decision to line up with the pods, hanging upside down as the motorboat swept by, and my perfect roll up.

"People can do the most incredible things when there's an emergency," Harvey said. "That's what it sounds like. The pods appeared, and you knew exactly what to do."

"Now it feels like somebody else did all of that."

Spruce Harbor was as peaceful as we left it. Harvey offered to take me home, but I insisted the few miles would be no problem. In my driveway, I filled a barrel with water and plopped the salty, sand-covered gear into it. I grabbed a few oatmeal cookies off the kitchen counter, trudged up the stairs, stripped off my fleece pants and shirt, and glanced at the clock next to the bed. Midnight.

I'd been gone seven hours. It felt like two days.

I padded into the bathroom, turned on the shower, and stepped in. The blessing of hot water tumbling over my head, shoulders, and the rest of my body almost made me cry.

In flannel pajamas, I fell into bed and within seconds drifted off. It was not my most peaceful night. In one dream, ice cubes bobbed around me in a wild ocean. In another, Ted chased me in a motorboat, then morphed into Seymour.

I woke to a bright April day, pushed aside the bedcovers, slid my feet into shearling slippers, and made my way downstairs. Dreamlike images of the Sunnyside escapade drifted through my mind. It was hard to believe it really happened. I pulled open the fridge door. The water samples sat on the shelf below the two-percent milk.

Tea was the right thing—a strong pot of English Breakfast. Holding a steaming mug in two hands, I rested my elbows on the café table and took in the view stretching beyond the deck. The indigo sea was calm, and I needed that. There was a lot to think about.

First of all, I'd been very lucky to escape disaster the night before. Clearly, I was too cavalier and misjudged the possibility of danger. My lesson was a well-known idiom. Look before you leap.

I also had to consider consequences of my escapade. Mainly, there was Ted. I'd obviously misjudged him in a big way. He could easily guess who he chased. After all, he knew I visited Sunnyside and was an avid kayaker.

Given that, I couldn't imagine how he'd act around me now. The loss of his friendship was really too bad. I'd thought he was such a great guy. Well, I'd been badly burned before by someone I trusted who turned out to be far from great.

Yes, look before you leap.

The tea's caffeine kicked in. Wait a minute. *Ted* told me about Frank's genetic engineering work. Maybe he tried to bait me for a reason I couldn't imagine. Damn. It was a mess.

A ringing phone startled me. The landline. I slid off the stool and reached for the receiver.

Angelo said, "I waited to call you since you needed the extra sleep. Connor and I want to hear what happened." He sounded good, upbeat. With my adventure safely over, Angelo was back to his normal self.

"Why don't you come over here for dinner for a change? You and Connor—and I'll ask Harvey too. Can you make it tonight?"

"That would be lovely, dear," he said. I smiled at his Maine pronunciation of dear—"deah."

I took a quick trip to my office to pick up some work. Since it was Sunday, maybe Ted wouldn't be around.

He was.

"Busy?" he asked, appearing in my office doorway.

My mouth flew wide open. I clamped it shut. "Ah—well—no."

Leaning against the doorframe, Ted was relaxed and all smiles.

It seemed like I'd stepped into a different dimension. How could Ted stand there looking like nothing happened? Had I imagined him at Sunnyside?

Pay attention to the data. You're a scientist. You saw him there. On the pier.

Ted looked really good. His tan—remnant of the Caribbean research trip—had faded some, and a lock of dirty blond hair fell over one eye. My hand twitched, and I wanted to push it off his forehead.

"Want to catch lunch and a beer over at the Absolute Bearing?" Ted asked.

I raised an eyebrow. The Absolute, as locals called it, isn't a bar where you drink beer and watch sports. It's an upscale tavern attached to an expensive restaurant.

But this man nearly ran over a kayaker. No way should I even consider having lunch with him.

Ted said, "If this isn't good for you, maybe we can do it another time?"

On the other hand, I might learn something if we had lunch. Intense desire to know what the hell was going on won.

Before I could stop myself, "Okay" came out of my mouth. Ted beamed.

"Terrific. Meet you there."

Halfway to the other side of the harbor I pulled over. This was nuts. I was going to lunch with a crazed man who probably knew I'd paddled the kayak he chased. But if that were the case, it didn't make sense that he was so nice to me now.

"You'll be in a public place," I said to myself. With a sigh and a look to the heavens, I pulled out onto the road.

In the Absolute Bearing lot, I sat in my parked car again. *You can still turn around. Make up some excuse.*

The restaurant entrance was thirty feet away. Ted was probably already inside. If I left, I'd drive myself bananas wondering what he was up to.

I stepped out of the car.

Deep breath. Act like nothing happened and see how it goes.

The Absolute sits right on the water. I walked quickly through a chilly crosswind, pulled open the massive front door—something off an old ship—and slipped inside. Ted greeted me, and we were seated at a window table. We barely had time for small talk when the waitress brought our meals—local beer and a bowl of chili for Ted and a glass of Sauvignon Blanc and Caesar salad for me.

Since this was a fishing expedition, I decided to go whole hog. "Sauvignon Blanc's Harvey's favorite white wine."

"Right. You two are good buds."

Empty hook.

Ted laughed.

"What's so funny?"

"Oh, I'm not sure what to talk with you about, unless it's algae or something."

I went for the obvious. "How 'bout where you grew up and went to school?"

Ted took a hefty spoonful of chili and a slug of beer. He wiped his mouth. "Kansas. Couldn't have been farther from the coast. But those amazing Cousteau shows on TV fascinated me when I was a kid. So I went east—University of Maine for college and grad school. What about you?"

I put down my fork and met his eye. "Born in Maine, but inland. I'm an only child and was young when we moved here. My father was offered a job at the new MOI. Bowdoin for college, grad school at MIT, my post-doc here." I steered the topic toward scientific fraud to see if Ted would nibble. "Did you read the story about that South Korean scientist? The one who falsified stem cell research? He said he'd cloned human embryos to get stem cells from them. You know, to cure diseases like Alzheimer's and Parkinson's."

Ted put down his beer glass and didn't miss a beat. "Yeah. He was fired from his university a while back."

I went deeper. "All we have as scientists is our reputation for integrity. We collect data, ask questions, and try to answer them as best we can. If our colleagues suspect dishonesty in how we do any of this, we've lost everything."

Ted's demeanor didn't give away a thing. He nodded. "I always stress this to my grad students."

I was stumped. Was Ted a talented actor?

"Ah. Did you ever do any acting when you were younger?"

This took him by surprise, but he recovered quickly. Grinning, he said, "How'd you know? In high school, I played Harold Hill, the Music Man."

I nearly groaned. Harold Hill, the con man.

Ted furrowed his brow. "Are you all right?"

"Huh? Sorry. Must be tired."

"Ah. You haven't mentioned Sunnyside. You must be curious to know more about the place."

Now, I was utterly confused. Given last night's events, Sunnyside was not something I wanted to discuss.

"What?" I said. "Actually, I haven't had time to think about it. You know. So busy."

I steered the conversation to likes and dislikes—safe territory once more. We both liked the Red Sox, blues music, and scuba diving but disagreed on football and Bonanza reruns (Ted's favorites) and Italian opera and cats (mine).

Driving home, my mind was spinning. I hadn't detected a hint of the nutcase who chased my kayak. On top of that, Ted wanted to talk about Sunnyside. Maybe he guessed who he'd chased and was trying to push *my* buttons.

On the other hand, as much as I hated to admit it, Ted was great company. It was amazing and perplexing how relaxed I felt. For the most part, the conversation was fun and easy going, like spending time with one of my girlfriends.

I sighed and turned in to my driveway.

By seven, Angelo, Connor, and Harvey had gathered around my kitchen table. Harvey wore a new dress—a fitted sheath in dark green—and looked stunning. I'd pulled on clean jeans and a red sweater and called it done.

Angelo and Connor helped themselves to tomato and basil bruschetta and told Harvey about the day's fishing trip. Harvey leaned over with her napkin and caught a drip of oil from Angelo's chin. Laughing, he rubbed his chin and thanked her.

So different from one another, they shared devotion to the natural world and to those closest to them. A rush of warmth flowed through me.

Dinner was a group effort. Connor grilled fish, Angelo made salad, Harvey set the table, and I put together apple crisp. My dining room faces seaward, and as we began our meal at twilight, we could see beach rose bushes moving about in the wind.

I updated Angelo and Connor about Sunnyside. Later, I wanted their reaction—and Harvey's—to an outrageous new idea.

We passed around grilled fish, hot rolls, and salad. Between bites, I narrated the story of my escapade step by step—from the drysuit struggle to my paddle back to Harvey. I slid over the scary stuff.

"She was one tired puppy when she landed," Harvey added.

As I spoke, Angelo's expression changed from interest to disbelief and dismay. He leaned against the table to massage his temples.

"I should never have agreed to your plan. My god. Who'd think someone growing algae would race after you with a motorboat?" He coughed and sat back in his chair.

Connor frowned. "Mara, there's something you haven't thought about."

"Yes?"

"It's very hard to identify people from a distance, especially if you're anxious."

Harvey picked right up on that. "Connor's right."

I looked away and pictured the scene up on the Sunnyside pier. "So I've heard. The docks were well lit, and I was really close. It *was* Ted. I'm sure. And it makes sense he'd be there. He studies algae, after all." I added, "Actually, there's more to tell." I looked at each of them and settled on Harvey. "Harvey doesn't even know this yet. Ted invited me for lunch today."

Harvey said, "And—?"

"I went."

She and Angelo mirrored each other—eyes wide, mouths open.

But Connor just chuckled. "You're quite the spy, Mara. First, your kayak caper. Now drinks with the enemy."

Harvey glanced at Connor. "I suppose that's one way to look at it. What happened? Did you learn anything?"

"He grew up in Kansas and likes football."

"That's it?"

I shrugged. "As far as connecting him to my kayak caper, as Connor calls it, nothing. I even asked his opinion about scientific fraud. He answered like any honest scientist would. It was a normal chat with a colleague."

Harvey looked like she was about to say something when Angelo stood up abruptly. "Could we have dessert and finish talking about this then?"

I bit my lip. "Sure."

Harvey and Connor helped clear the table while Angelo went out to his car. We all settled in the living room with coffee and apple crisp. Angelo slid his half-eaten dessert across the coffee table. He held up a folded newspaper. The *Boston Globe*.

He directed his words to me. "It's another anniversary of the Boston Marathon bombing. There's a piece here about the younger brother, reactions of people who knew him."

"Yes?"

"They all say the same thing. It's impossible to believe the young man they knew—a student, friend of their nephew's—killed and maimed all those people." I opened my mouth to respond, but Angelo stopped me. "I'm not finished, Mara. People who live next door to serial murderers say things like 'he's such a nice guy' or 'he's the last person who'd do that.'"

"Wait a minute. That's a little extreme. I didn't walk down a dark alley with a serial killer. I had a drink in a public place."

I planted my cup on to the coffee table so hard it spilled, sat back, and crossed my arms. Naturally, Angelo was concerned. But I wanted him to sympathize with my point of view. Angelo's lips were pursed, and his arms were across his chest as well.

"Hear me out," I said, hands in front, palms up. "This is driving me crazy, trying to square an insane man in a motorboat with the Ted on our cruise, the guy I had lunch with."

Connor leaned forward. "But what can you do? You can't walk into the police department and report a crazy man who tried to mow you down with his boat. They'd wonder what *you* were doing out there at night and assume the guy was a security guard."

"Two things. First, I can get the algae samples analyzed. If we guessed right, the *Chloronella* are being fertilized with commercially made nutrients. The isotope data will show that."

Connor leaned back and slid a little closer to Harvey.

Harvey spoke up. "But even if you find out their claim is a hoax, how does that help?"

"I can tell Sarah this probably was the fraud Peter discovered. Right now, anything that can give her some closure is huge. And John Hamilton deserves to know what's really going on, of course."

Angelo cleared his throat—loudly—and said, "Actually, I'd like to go back to first base here."

Angelo could be firm but was rarely edgy. The three of us stared at him.

"Tell me, what does Ted, isotopes, and the rest of it have to do with Peter's death? Isn't that why you started working on this in the first place?"

Harvey and Connor's heads swung back in my direction.

"Okay," I said. "I can't link scientific fraud at Sunnyside to the *Intrepid* disaster, but there might be a connection. Someone, Ted, had to be desperate to nearly kill a trespasser in a kayak. There's a big secret there. Maybe it's something worth killing for—twice."

Harvey scrunched up her mouth like she does when she's not convinced. "Then somebody went to an awful lot of trouble to kill Peter on a ship."

"And," Angelo said, "*you* were scheduled to supervise that buoy deployment, not Peter."

I ran my fingers back through my ponytail. "Everything you say makes sense. I just don't know."

All three were pensive, sifting through the bits and pieces.

Connor walked to the fireplace and turned to face us. Excited, his blue eyes shone. He was full steam ahead.

"Let's go with that, what you said. There's a connection between what happened on the ship and the guy chasing you. How can you figure out what that is?"

Harvey looked up at Connor with interest.

Grateful for his zeal, I smiled. "I've been trying to work that out and think I've hit on a plan."

In an exaggerated Irish brogue Connor said, "Well, fill us in, lass."

"It involves Ted."

"Please, Mara," Angelo said. "If the man is dangerous, it's not safe for you to have anything to do with him."

"If I talk with him, it will be in a public place again. He emailed a little while ago and invited me to meet him tomorrow night at the Captain's Shack Restaurant." I turned to Connor. "You could even be there in a corner table, you know, to watch him."

Connor picked right up on that. "I could take Harvey out to eat at the same time. Ted doesn't know me. So even if he notices us, he'll think we're out on a date." He grinned at Harvey, a goofy grin.

I couldn't read Harvey's reaction.

Angelo threw up his hands. "But I still don't understand why you want to talk with Ted."

"Two invitations right after the kayak trip? Isn't that a little much?"

"Like he's worried you spotted him at the algae place, and he's trying to make it seem impossible?" Connor asked.

"Maybe. It's all too weird. Again, I want to know, have to know, what's going on with him."

"So if you go to dinner, what will you say?" asked Harvey.

"Just tell him I kayaked to Sunnyside and saw him and gauge his reaction."

Angelo's sigh was so pronounced we could have recorded it for a "sounds people make" website. For the next ten minutes, he worked hard to get me to change my mind. His main argument was a good one. Ted was someone I'd want to work with. By accusing him, I could lose an important colleague forever.

Through all of this, Harvey and Connor sat close to one another on the couch, quietly conspiring.

Finally, Angelo held out both his hands, palms down. An Italian gesture of defeat. "My god, you're obstinate." He dropped back into his armchair.

Harvey said, "See what you both think of this plan, Mara. You'll accept Ted's invitation and make up an excuse to meet him at the restaurant. Connor and I will go early so we're already seated when you and Ted arrive. A table out of the way. The restaurant's not that big, so we can keep an eye on you the whole time."

"I like it. That's the plan."

I got up and lit a match under crumpled newspapers in the fireplace. The white birch logs were soon ablaze. For the rest of the evening we didn't talk about isotopes, algae, or crazy people. With Connor and Harvey's assurances, Angelo even warmed a tad to the new idea.

Connor kept sneaking looks at Harvey in her green dress. He didn't have a chance if Harvey and Ted were a couple.

And if they were, she sure was hiding it well—and exposing Ted might well jeopardize my closest friendship.

15

I SAT AT MY DESK early the next morning plugging away at the grant proposal. It was a good distraction. I itched to look at the buoys' temperature data, but two days remained in my self-imposed refrainment.

For the proposal, I had to trust Gordy's judgment. The lobstermen saw me as "diff'rent," so they might be willing to work with global warming scientists, if I picked the right people. After all, the NSF wanted to see creative ways contrary groups like the Fishermen's League and scientists might cooperate. Using my whiteboard, I jotted down names of scientists I might ask, crossed out some, and added others. Eventually I had a good list and needed to clear my brain with a run.

I jogged down the MOI driveway, along the wharf to the other side of Spruce Harbor, and back—four miles in all. For good measure, I trotted up the stairs to my floor.

I squinted. It looked like someone had dropped a red scarf outside my door.

I got closer, ran to the red thing, and fell on to my knees. "No. No. No. You poor baby. No."

Homer was limp and drying in a small puddle of water. I cradled him in my arms. He didn't move.

As quickly as I could, I carried Homer—not easy with such a big lobster—to the elevator, down to the basement, and over to his tank. I slowly lowered him into the water. He sank to the bottom, motionless. I pressed my forehead against the glass window. The salty water running down my cheeks dripped onto the salt-encrusted floor.

For the next hour, I sat on a stool in front of Homer's tank and waited, peering in to see if he'd move. I was alone. Just me, the rumble of seawater, and a lifeless lobster.

I felt so very sorry for the suffering of this peaceful animal. Then I got angry. The longer I stayed, the angrier I got. Finally, I was absolutely furious.

I paced back and forth in front of Homer's tank. Someone had done this to get at me. They'd taken this innocent creature and dumped him on the floor for me to find. I wondered how long Homer lay drying out on my threshold.

One of Homer's legs moved a bit. Pressing my forehead against the window I whispered, "Homer, if you rally, I'll find out who did this to you."

Sluggishly, the lobster moved his eight walking legs and rotated his body so that his head was just on the other side of the glass from mine.

We had a deal.

The afternoon was shot workwise, and I had to talk to someone about what'd happened to Homer. When I felt sure he was okay, I went back upstairs. Fortunately, Harvey was in her office finishing up with a grad student.

Slipping into the chair the student had just left, I bent over and rubbed my neck.

"What's happened?"

I told her.

Harvey gasped. "That lovely animal. It's cruel. What kind of a sick person would do that?"

"Right. And why?"

"What're you thinking?"

"It's a message."

"Keep going."

"Here's my guess. Whoever did this could've killed the lobster, but didn't. There was water on the floor."

"Which means—?"

"Homer hadn't been there that long. Maybe someone watched me leave for my run, knew I'd go back to my office, and left the lobster there. You know, as a warning. Like stay away, or else."

"Someone in our building?"

I nodded.

"Mara, Ted didn't do any of this."

"Guess I'll find out."

On the way home, I ran errands. Back at my house, I showered and got dressed for dinner.

As much as I didn't want to admit it, I looked forward to being with Ted again. No doubt about it, he *was* a pleasure to talk with about almost anything. We shared so much. He had a great sense of humor.

Maybe he was Dr. Jekyll and Mr. Hyde—or just a great actor. Or, he might enjoy dangling me on a string like a puppet. Then there was Homer. Ted was a biologist. Could he possibly have it in him to take the lobster out of his tank and dump him on the floor outside my office?

There was something else—something I didn't want to own up to. I had been blind to a good-looking guy's dark side before.

Tall, dark, handsome, Davie was dark in ways I hadn't seen. While we were in what I assumed was a committed relationship, he had three other sex partners. Three. Looking back, I was amazed at my own naiveté. I missed the obvious signs—the unexplained trips, the many nights he told me he had to work late, the lack of care and attention I deserved as his supposed other half. But when he ran those chestnut-brown eyes across my body, well…

I brushed my hair until the red highlights shone and looked into the mirror to get my barrettes right. Maybe I was looking at the face of a fool.

The Captain's Shack is on the north end of Spruce Harbor. It serves delicious seafood and isn't expensive, making it a favorite with locals. I pulled into the parking lot. There were so many cars there already I had to leave my car at the lot's edge by the water.

I walked slowly toward the entrance and inhaled cool, salty air. I yanked the door open and stepped into the buzz of a busy restaurant. Ted sat at the far end of the long wooden bar. I skirted a klatch of twenty-somethings. Sipping her aperitif, one had fixed her eye on Ted.

I slid onto the bar stool next to him. He wore a light blue shirt open at the neck, an indigo sweater vest that set off his eyes, and tan cords. That same unruly bit of hair fell over his forehead.

"Sauvignon Blanc?"

"Perfect."

He ordered the wine and turned back to me. "You look terrific."

I blush easily, and could feel the heat come to my cheeks. Like a little kid, I looked down at my clothes—black velvet jeans, black ankle boots, a red cashmere sweater. Decidedly nicer than my usual attire.

Glancing up at him I said, "Um, thanks."

Thank goodness, the maître d' called Ted's name and led us to a table with a view of the bay. I sat down and looked around. Harvey and Connor were at a table in a back corner. She winked and returned to what looked like a lively conversation. Near Harvey, a camera flashed. If the photographer was from the *Spruce Harbor Gazette*, the annual spring byline—"Locals enjoy last quiet before tourists arrive"—would be in the next issue.

Over dinner Ted and I chatted about upcoming meetings and some papers we'd read, but I was distracted and lost the train of the conversation. I was trying to figure out how to bring up what was foremost on my mind without making Ted angry. But of course he'd get angry.

Ted made it easy. "Mara, I'm not the most sensitive guy, but it's pretty clear something's bothering you. What's up?"

I shoved my plate to the side and leaned on the table. "Okay, I'm going to tell you. Promise you'll listen all the way through. I don't know what you're going to say, but I'm not good at hiding things and want to get this out."

Ted crossed his arms and leaned back. "Fire."

"Two nights ago, I paddled out to the Sunnyside piers."

Ted raised an eyebrow.

"To find out if the genetic engineering was a hoax, I collected water samples along a transection from the shore to the *Chloronella* tanks. Ted, I saw you there on the pier. And heard you add fertilizer to those tanks."

Ted's expression went from surprise to outrage in a half-second. His tone was flat and cold. "Mara, that evening I drove to Boston to visit my parents. I wasn't even in the state of Maine."

Nothing was going to stop me now. "It was dark, and I couldn't see perfectly. But I was very close and the lights were bright. The guy on that pier looked just like you. Face, size, hair. All exactly like yours."

Ted frowned and shook his head. His response was low, almost a whisper. "It was Frank."

"Frank?" I echoed.

A rush of realization came over me. Frank, the childhood buddy who people took as Ted's brother. Frank was a scientist too, and he also lived in Maine. In my rush to identify my pursuer, I'd overlooked an obvious explanation.

"Frank?" I repeated. The information slowly sank in. I groaned. "Oh lord, it wasn't you. It was Frank."

Ted's face was grim, hard.

I tried to backpedal. "Ted, I don't know what to say. Yes, I did think it was you. But part of me *didn't* believe that, which is why I'm here with you now." Elbow on the table, I rubbed my forehead.

Harvey shot me an "Are you okay?" look. I nodded my head ever so slightly.

Ted stared at me.

He spoke, his words measured. "Mara, the very idea of lying about a project is an abomination to me, just as it is for you. Surely, you know that?"

I searched his face. "I wanted to believe it. But it was undeniable evidence. What I saw with my own eyes." Looking down at my hands I added, "I should've considered other explanations instead of getting carried away." It sounded pretty lame.

"I don't get why the hell you were out there, but at the moment I don't give a goddamn." He marched away from the table, settled the bill with our waiter, and left. I stared at the chair across from me, feeling like a complete idiot.

Harvey slipped into Ted's chair. Connor stood behind her.

She reached over to touch my arm. "What happened, Mara?"

I groaned. "I told him the whole thing. He got angry and left. It couldn't have been him. He was visiting his parents. Can't blame him. I'd be furious."

I covered my face with my hands and shook my head. What I'd done was just sinking in.

Connor asked, "So who was on the pier?"

I looked up at him. "It was Frank, Ted's childhood friend, who works at Sunnyside. They look a lot alike. He'd told me that, but I forgot."

Harvey whispered, "Of course."

"Look, it was an honest mistake," Connor said. "I'll bet your boyfriend will understand that in a few days. Give him time."

I cringed at the "boyfriend" label but didn't say anything. Connor was just being kind.

"Let's get out of here," he added.

Outside, Harvey asked, "Where'd you park?"

I pointed to the left.

She gestured in the other direction. "We're over there. Want us to walk you to your car?"

"No. I'll be fine."

She hugged me. "Look, I'll see you tomorrow. We can talk about this then. Let's have breakfast together at the Neap Tide, okay?"

I nodded and watched Harvey and Connor turn the corner. The wind had come up. I rubbed my arms and walked quickly toward my car. It sat in a dim pool of light beneath one of the few lamps in a far corner of the lot.

I skirted potholes in the now nearly empty lot and rummaged in my purse for my keys. I reached the car, but they were still buried. "Too much junk in this bag," I mumbled.

That's when it registered. Feet on gravel. Fast moving footsteps.

Desperately, my fingers groped through pens, glasses, crumpled receipts—before they touched the hard metal of car keys.

I yanked up the keys, dropped my purse. Fumbling for the right button, I glanced back. Bared teeth flashed in the yellow light.

This was a madman.

He growled and was on me in an instant. He grabbed my shoulders and slammed me hard against the car door.

Somehow, I managed to twist around and knee him in the groin. He groaned, loosening his grip.

I tried to scramble away. He lunged for my shoulders again and slammed me to the ground. I blocked the fall with my hands and slid forward. Gravel dug into my palms. I knew what it meant to see stars.

The man jumped on top of me with spine-snapping force.

I squirmed and tried to call out, but his weight crushed me. He grabbed my hair and yanked back hard. My skull smashed into neck bones, the screaming pain of hair ripped out by the roots red hot.

He hissed, "You think you're smart, Dr. Nosy. You and your kayak." He pulled harder.

Blessed Mother of God. This is how I'm going to die.

Electric pain, panic, and terror ran through me like water in a swirling flood.

Sound faded, my body spun, and darkness began to overtake me.

Suddenly, the torrent stopped.

The crushing weight was gone.

My head fell forward. My eyes popped open. I could breathe.

Irish brogue. "That's it, Paddy! Get the hell off the lady. Let's see your face before I punch you to bloody Timbuktu!"

Gasping, I rolled over on to my back and squinted. In a dull pool of light, Connor held the spitting image of Ted a foot off the ground. Like a puppet, the guy hung by the shoulders of his jacket as Connor shook the hell out of him. Behind Connor, Harvey frantically spoke into her phone.

A flash of a camera went off as the photographer caught Connor dangling Frank in the air for all to see—in the next edition of the *Spruce Harbor Gazette*.

The real Ted ran out of the darkness and over to me. I sat with Connor's coat wrapped around my shoulders. Harvey knelt next to me.

Ted said, "I was driving out and saw a flash. What the hell's going on?"

I tried to raise my head and winced.

Harvey answered for me. "That man attacked her. Mara was just about to get into her car when he threw her to the ground and tried to strangle her."

Ted looked down at me, over at Frank. "Jesus."

He walked to Frank's side and tried to talk to him. Frank twisted his head this way and that, mumbling senselessly to the air. Connor stood guard so that Frank didn't try to escape.

Harvey and I glanced at Ted, Frank, back at Ted. They could have easily passed for twins, except that Ted's good looks had given way to a hardness in Frank.

I mumbled, "Jekyll and Hyde."

The scream of sirens filled the parking lot. By the time the EMTs ran over to me, my whole body shook. They checked my vitals and said I wasn't in immediate danger. When I complained about sharp pain in my neck, they decided I needed medical attention at the hospital.

A policeman asked a few questions—my identification, if I could briefly describe the assault, who attacked me. To that, I motioned toward Frank. The officer asked if I knew why Frank assaulted me. I hesitated. "Um, water samples."

"Water samples?"

I blinked.

"We'll deal with this tomorrow," he said.

The medics strapped me to a stretcher and slid me into an ambulance. My head throbbed through every turn but since the nearest emergency room wasn't far from the Spruce Harbor waterfront, I was rolled out of the ambulance and down a hallway in minutes. A clamor of voices hollered orders and a pair of strong hands moved my head from side to side as an IV poked my arm. I awoke with a bandaged head to news that X-rays showed no bone damage and Harvey, Connor, and Angelo were waiting to take me home.

The next morning I woke with a start. Blinking, I looked at the ceiling and recognized a light fixture. I was home in my own bed. Had last night actually happened? I turned my head toward the window and grimaced with pain. That was very real. My back muscles screamed in protest.

Someone moved around in the kitchen below me, then climbed up the stairs. Harvey poked her head into the bedroom.

"Here's some coffee, sweetie." Harvey put a steaming mug on my nightstand. Gingerly and by inches, I pushed myself to a seated position as she pulled my pillows up against the headboard.

I leaned back with a sigh and gratefully accepted the coffee. Harvey sat on the edge of the bed. I was in my own bedroom—warm and in one piece. I had Connor and Harvey to thank for that.

"I'd give you a hug, but it hurts too much to lean over. I remember medics, a policeman, the emergency room, you and Connor walking me into the house. The rest is a blank."

Harvey said she got me ready for bed, stayed the night, and checked a couple of times to make sure I was okay. Connor left around three in the morning and would tell us what happened to Frank.

"Ted?" I said tentatively.

"Ted tried to talk to Frank for a few minutes. After that, I don't know."

I shifted in the bed, winced, and rubbed my neck.

"How do you feel?"

"My palms and knees sting where the gravel dug in. My arms are really sore, and my neck hurts like the devil, but that will all pass."

Harvey patted my leg and said she'd return in a few hours. With a sigh, I slid back down under the covers and instantly fell asleep.

By late morning, I was dressed and downstairs making brunch, slowly. Work was out of the question, and I'd cancelled my meetings. That gave me free time to clear my head.

Angelo phoned at noon. I barely recalled seeing him at the hospital. Connor had called him several times, so he was well informed.

"Guess you were right to be apprehensive," I said.

"I was completely off base about who was dangerous. But what matters is that you're all right."

Angelo was so decent. No hint of "I told you so."

"I love you, Angelo."

"And I love you too, dear."

The same policeman who'd talked to me in the restaurant parking lot dropped by to finish his set of questions. Again, I tried to explain why Frank attacked me.

He looked down at his scribbles. "See if I've got this straight. On Saturday night, you paddled your kayak from the beach to a pier at the Sunnyside place in Winslow Bay. You got water samples along the way, which you put into little vials. You didn't get out of your boat or land on the pier. Frank Lamark saw you and took you for a trespasser. You paddled away from the pier and he chased you with a motorboat."

"That's right. I was sure he was going to run me over. To escape, I rolled the kayak and let him pass by. I rolled back up and paddled away."

The cop eyed me. "Yeah. Then, last evening you had dinner at the Captain's Shack with, um."

"Ted McKnight. Frank's childhood friend."

"Right, Ted McKnight. After dinner, you walked to your car alone because Ted left earlier. Frank Lamark assaulted you as you reached your car at the far end of the lot. You believe he'd discovered you were the kayaker he chased."

"All that's correct."

The cop closed his notebook. "A judge has seen Frank Lamark and requested a psychiatric examination. Frank's parents have posted bail, and he's staying with them. Do you want to press charges now or wait until Frank has been diagnosed?"

"I'll wait."

After the officer left, I was desperate for fresh air. At the pace of an old lady, I walked into the bright day and gingerly took little steps along the footpath to the beach. At the water's edge, I stopped to drink in the scene. A light wind rippled the surface of sapphire water. Gulls cried hoarsely. Far off, a lobster boat droned its way back to Spruce Harbor.

I picked my way along. As always, the little things delighted me. The tide was low, and I mimicked a gull's wandering footprints for a bit. Their tracks alternate like ours, and I tried to match the pigeon-toed stride.

Sharing a gait with gulls—salt-laden wind in my face—this was joy. Nobody could be happier to be alive.

When I got back, Harvey's car was in my driveway.

"You look pretty chipper." She breezed into the kitchen and plopped a grocery bag on to the counter.

I followed at a slower pace. "You said you'd bring a casserole. This looks like food for a week."

From the bag Harvey pulled out ice cream, goat cheese, English crackers, grapes, lettuce, tomatoes, avocado, lasagna. "All things you like. You deserve some treats. How're you feeling?"

"Like I did sumo wrestling—with no training. But I'm loosening up. I'll make some tea."

Five minutes later, we sat across from each other at the kitchen table, hands cradling mugs.

"There's one thing I don't get," Harvey said. "How did Frank know who you were? He didn't see who was in the kayak, after all."

I sipped my tea. "Been thinking about that. When I visited Sunnyside, somebody was out on the piers. His back was turned, but he had blond hair so it probably was Frank. Maybe he was curious about my visit and looked me up online. My photo's right on the MOI website, of course."

"That makes sense, especially if he's hiding something. But how would he connect you with the kayaker?"

I groaned. "On the website you can list your non-academic interests."

"You didn't."

"I noted that sea kayaking is my favorite sport and that I extensively explore Maine's coastline with my kayak."

"I suppose it's not much of a leap. But why attack you in the parking lot?"

"Remember, the guy's crazy. You heard him mumbling nonsense. When I left, my car was in a deserted part of the lot. He could've gotten away with—" I rubbed my neck again.

"You know, Mara, the time to quit is before you look back and wish you had."

There was nothing I could say to that.

I changed the subject. "Tomorrow the isotope results should come in."

Harvey blew on her tea. "And?"

"If we're right, I should tell Hamilton that Frank's adding fertilizer. God, what a mess."

"Don't you think he'd know?"

"He leaves all algae-related details to Frank."

"What will Hamilton's reaction be?"

I shrugged, then winced. "He won't like that I kayaked near his piers, but I didn't do anything illegal. I hope he focuses on what the numbers tell us."

"And Ted?"

I put my cup down and sighed. "Connor's right. I have to give Ted time. Either he's going to see how I could've confused him with Frank and forgive me, or he's not. Nothing I can do about it now."

We hung out in my living room until late afternoon. Mostly we read. Harvey tackled student papers. I started a book about an unheralded female cartographer whose underwater maps set the stage for continental drift research. As the sun faded and wind picked up, I got a fire going. The room was soon cozy warm.

We had been quietly reading when I said, "Do you get lonely? I mean, do you want a guy?"

COLD BLOOD, HOT SEA

Harvey took off her reading glasses. "Where did that come from?"

"I don't know, we've both lived alone for years now."

"The men around here don't interest me. Most scientists are married, and the locals are, well, too rough."

Harvey had never admitted this, but it didn't surprise me. She'd grown up in a wealthy family and attended an expensive boarding school and college. Even though she loved hunting, her background set her apart from most men who lived in Spruce Harbor.

I didn't ask about Ted. If they were a couple, I wanted Harvey to tell me. It was her private life, after all. But I figured a prompt wouldn't hurt. "So there's no MOI scientist who interests you at all?"

She shrugged.

"What about Connor?"

"He's a good guy, very funny. We talk about hunting, and I really enjoyed our dinner. Until what happened, of course. But you know, he's not my usual type."

"He'd be a great guy to have around," I teased.

"You come home to an empty house, too."

My usual retort—"and I don't have to pick up anyone's socks"—felt hollow, so I didn't say it.

Harvey, sporting the hint of a cat's grin, was mum.

Harvey left and I went back into the living room, sat before the fire, and stared at it, savoring the tranquility. Lord, it was delightful. The flames died down. I got up to stretch my legs.

Enough pampering. It was time to be candid with myself. Harvey's words came back to me.

The time to quit is before you look back and wish you had.

Maybe it *was* time to quit.

I paced the room, outlining the things on my mind.

One. I held up my forefinger. I'd nearly ended up dead—twice—because I insisted on getting water samples and wouldn't let that go.

Two fingers. Ted. Talk about being blind. By paying attention to my science smarts, I would've admitted the lighting was unnatural and whole situation weird. Like Harvey said, it would be easy to be wrong about who was on that pier.

The lesson was obvious, and in the last few days three people—Angelo, Harvey, and Ted—had commented on one of my greatest shortcomings. Stubborn, pig-headed, too sure of myself.

Third finger. Honesty. I thought a good deal about being truthful in my profession but hardly at all about owning up to my faults honestly.

Okay. I'd be honest with myself.

But I wouldn't quit.

Suddenly, I was dog-tired. I stepped outside on to my deck. The moon hadn't risen yet, and the sky was littered with stars, some so bright they reflected off the glass ocean.

Peter died over a week ago and my progress to understand his death was still pitiful. I was letting Peter and Sarah down. I looked up, closed my eyes and asked for some help, and went back inside.

16

THE NEXT MORNING I WALKED into my office and picked up the ringing phone. It was a technician at the isotope laboratory.

"You said this was important. I'll email a spreadsheet with the data later, but should I fax the instrument's printout now?"

"Yes, please. Thank you so very much."

I wanted to sprint down to the first floor but was still sore and took the elevator. The door slid open. I spotted Seymour standing in the office doorway and sauntered toward him. I didn't want him to guess that something was up—something he'd stick his nose into.

Sliding by him, I said, "Mornin', Seymour." He grunted and returned to his conversation with the administrative assistant. That was good. By now, he'd probably know an MOI scientist was attacked, but apparently my name hadn't been released.

When Seymour was out of sight, I quickened my step. In the copy room, two pieces of paper rolled out of the fax machine. I snatched them, leaned against the wall, and looked over the data.

Just what we suspected. The isotope signature indicated the source of nitrogen contamination in Winslow Bay was commercial fertilizer from the Sunnyside piers.

Assuming that was the case, I'd confirmed a biofuel hoax. Most likely, the hoax Peter had discovered.

Back in my office, I put the two printouts side by side on my desk to study the numbers. There was a quiet knock on my door.

"Come on in!"

I looked up. The data sheets slid onto the floor.

Ted stood in the doorway.

He pointed to the chair next to my desk. "May I sit down?"

I nodded, placed my palm on my chest, and waited for him to get settled.

He looked down at his hands, then up at me. His face was drawn, hair rumpled. "I want to talk about what happened."

I managed to get out, "Um, of course."

"But first. Are you all right? I mean, not hurt?"

"Sore, but I'll be okay."

His eyes, deep blue, flashed. "I was furious you'd think I could be dishonest like that. Lie about my research. Well, you know that. After I left the restaurant, I sat in my car trying to cool down, which took a while. I headed out of the lot. That's when I saw the commotion. The whole thing was a nightmare. You were on the ground, and Frank acted crazy. He'd never been anything close to that before. I felt confused and helpless."

I leaned forward and said quietly, "Ted, you don't have to—"

"Yes, I *do* have to. I want to tell you this.

"At home, I pulled out some old photos of me and Frank. Maybe I wanted to get back in touch with the old Frank. Anyway, I saw how much we looked alike. Like twins. People always said that, but the photos made me realize how true it was. So I could see how you might have thought that, that—"

My body relaxed as relief poured through me. I let go of the armrests I'd been gripping. Maybe we'd work through this chaos after all.

"Now, it's my turn. Yes, the man on the pier looked like you. But I got completely carried away—so sure I was right. Stupid, but I didn't think about alternatives."

I talked faster. "Looking back, well, I feel so—ah—completely foolish. A little voice told me something else was going on. That's why I agreed to have dinner with you. But I'm not the most tactful person and just blurt stuff out sometimes." I looked away and back at Ted. "I truly am so very sorry."

"Scientists aren't supposed to assume they're right."

"Yeah. We follow the evidence. I can do that with my research. But it's a whole lot harder when it's me." I bit my lip.

We stared at each other for a few seconds.

I said, "Speaking of evidence, can I show you something?"

Ted looked relieved to change the subject. "What is it?"

I picked up the printout and handed it to him. "Take a look. Isotope values from water samples I collected from my kayak. It's a transect from shore to right below the *Chloronella* tanks. That's what I was doing there."

He studied the numbers and let out a low whistle. "So you're thinking that fertilizer's coming from Sunnyside?"

"Yes. It appears the *Chloronella* are getting nitrogen from a commercial source. Harvey will run the water samples, and the nitrogen concentrations will help verify that."

He read through the data again. "If you're right, Frank's delusional or lying. He may've been researching how to insert the nif gene. But based on this, he hasn't succeeded."

I couldn't read Ted's response. "You don't seem surprised."

He looked up from the printout, started to speak, and shook his head. He said quietly, "Honestly, I don't know what to think about Frank. It's all so bizarre."

My heart went out to him. "This must be dreadful for you. Frank's ruined any future in science. At the moment he's obsessed, dangerous, and he's led John Hamilton on."

Ted stared out at the bay. "Yeah, I know."

I waited. Ted's face was taut.

"Speaking of Hamilton," I said.

Ted blinked and looked at me.

"He'll want to know about Frank's dishonesty. He's an ambitious businessman and not stupid. Maybe he's already figured it out for himself. You know, that Frank's behavior was odd. Maybe he sent out some samples himself."

Ted nodded.

"Do you know anything about John Hamilton?" I asked.

"First time I met him was on the ship. Why?"

"There's another possibility. Hamilton may be in on this. What if they both thought Frank had successfully engineered *Chloronella*, then realized he hadn't. It could've been Hamilton's idea to add fertilizer and buy time."

"I suppose that makes sense."

"Let's see how he reacts. In any case, I should show him the isotope data."

"I want to go with you."

"Why? I've met Hamilton. I'm fine on my own."

Ted walked to the window and looked out, his back toward me. "I know that. Don't misunderstand me." He turned. "Frank's almost a brother to me. I don't get what's going on but owe it to him to try to make things right."

Ted's eyes were deeper blue, as if pain had blackened them. How could I not say yes?

"Sure. Of course you can come."

I called Sunnyside. John Hamilton was free that afternoon. Saying he'd be back in a half-hour, Ted left, shutting the door softly behind him.

I stared at the closed door and ran my fingers through my hair. My conversation with Ted had been a roller coaster.

He'd forgiven me, understood my mistake. I should feel relieved with such a burden lifted. But pin-pricks of doubt soured that pleasure.

Ted had gotten over being angry very quickly. My accusations were horrendous—deceiving the scientific community and chasing me with a motorboat. If someone had accused me of such things, it'd be a long time before I'd be willing to even talk to them—if ever. Also, right after I explained I'd visit Hamilton in person, Ted said he wanted to go with me.

Damn, damn, damn. That nagging question, again.

Was something else going on here?

An hour later, I was in my car heading north toward Sunnyside once more. This time Ted was in the passenger seat. We'd passed through Spruce Harbor and were well on our way up Route 1 before he said a word.

"Won't Hamilton ask how you collected the samples—like if you went up on his pier?"

"Anyone can collect water from that bay. There's nothing illegal there, and they didn't post warning signs. Anyway, let's hope he's more interested in what the data show."

We reached Winslow Bay, and once more I followed the long driveway to Sunnyside's main entrance. We stepped into the lobby. The tile floor was so clean I hesitated to walk on it, the room large enough for a party. Big money. A scandal would make it much harder for Hamilton to get funding in the future.

John Hamilton greeted us at the top of the stairs. Everything about the man was brown—from his hair to the rumpled sports jacket, tan shirt, and khaki pants falling over scuffed brown shoes. He already looked worn-out, and my news might devastate him.

We followed Hamilton into his office and over to the chairs beside his picture window. I glanced down at the bubbling algae tanks. It was hard to believe a few days earlier Frank had chased me—and I'd decoyed myself as one of the pods bobbing out there.

Ted stepped forward. "John, I'm Ted McKnight. We met briefly on *Intrepid*, but I was in foul weather gear. You may not recognize me now."

Hamilton nodded and looked at his watch. "Georgina should be here by now."

Georgina, Hamilton's wife. I'd forgotten about her.

Hamilton settled in his leather chair and tossed an over-sized paperclip between his cupped hands. "She'll be along. What's this about?"

"To come right to the point, I know Frank claims to have genetically engineered an algae—that he's inserted the nif gene into *Chloronella*."

Hamilton raised an eyebrow and put down the paperclip.

I pulled the data sheets out of my bag and handed them to him. "But these data indicate Frank hasn't succeeded and has been adding a commercial fertilizer. That's probably why the algae are doing so well." I explained what the numbers on the printout meant. "Of course, you should send out your own samples. But the fertilizer signal is strong. I'm confident they'll show the same thing."

Frowning, Hamilton looked down at the data. "I don't understand. How do you know about this research—and how did you get the water samples?"

I glanced at Ted.

"My sources are confidential. I got the samples myself. I kayaked out and collected water along the way."

"You took a boat out to our docks?"

"I didn't trespass on your piers."

Hamilton dropped the printout on the coffee table. He looked at Ted. "You agree with Dr. Tusconi's conclusions?"

"I'm afraid so, John. Like Mara said, the isotope signal is very strong."

John Hamilton rubbed his eyes and ran his fingers through his thin hair. He walked over to the window and looked down at his domain.

Ted and I waited for him to digest the information we'd given him.

Finally, Hamilton turned to look at us. Shoulders sagging, face ashen, he looked older. He shook his head. "I had no idea this was going on."

I skipped over the boat chase but said, "You know, Frank

attacked me outside a restaurant the other night."

"Actually, I didn't know who he attacked. He's been in a psychiatric hospital down in Portland."

"He must've realized I'd paddled out here."

Hamilton turned back and put his hand on the window. His voice quivered. "I wanted it to be true. We first grew algae without nitrogen fertilizer. Algal biofuels, I wanted a big part in that. Frank's so bright and driven. And good with his hands."

Ted got up and stood next to him. "John, Frank is my closest childhood friend. That's why I'm here. He functioned well in the laboratory, but now he may live in a dream world where he's the savior-scientist. He might genuinely believe he genetically engineered *Chloronella*. I don't know."

Hamilton looked at Ted like he'd forgotten he was even there.

Ted added, "None of this is your fault."

Hamilton blew out a long breath. "Yes. But now. What in heaven's name do we do about this?"

Georgina swept into the room, high heels clacking against the tile. "Do about what?"

She spotted the data sheet on the coffee table and snatched it up. "What's this?"

Hamilton introduced us and explained what I'd discovered. She towered over him, and his hands shook as he pointed to the data.

Frowning, Georgina looked down at the printout and over at me. Her eyes narrowed, and she waved the sheet in the air. "What the hell do you think you're doing, *Dr.*"—she exaggerated the word—"Tusconi? However you got these numbers, you trespassed in a restricted area. We could take you to court for this!"

"But I didn't trespass and thought you'd want to know—"

She cut me off. "We can do our own analyses and don't need nosy do-good scientists butting in where they don't belong."

I glanced over at Hamilton. He bit his lip—then shifted his eyes from me, to Ted and his wife, back at me.

I stood and faced Georgina. She crossed her arms and glared, head cocked to one side. Her hair, pulled back tightly, exposed a taut forehead, pencil thin eyebrows, and expertly brushed eyelashes. Bright red lips were jarring on her thin, pale face.

I looked at her straight on. "I was trying to be helpful—"

"*Helpful?*"

"I suggest you run your own samples."

She tossed her head. "We will. *And* we're not finished with you."

"Oh, yes you are." I headed for the door with Ted right behind me.

At the top of the stairs, I turned and looked back. Through his picture window John Hamilton looked down at the algae tanks. He was a decent guy who had trusted a scientist who'd betrayed him. Another of Frank's victims.

With a nasty, controlling wife.

Back in the driver's seat, I snapped my seat belt into place. "Whew! She's a piece of work."

"Sure is. I feel sorry for the man, having a wife like that."

I leaned back and tried to let go of tension before driving off. "It's good that's over—and that we did it. He cares a great deal about biofuel research and deserves to know what's going on. His wife clearly likes to be in charge, and her first concern is success of the business. That doesn't exonerate her, but she's got to feel threatened by our news."

"And won't admit it."

"Right. Think she'll get the water analyzed?"

"She's aggressive and smart. Sure. She wants to know what's going on, just not from you."

"This is really bad for John, isn't it? He counts on the *Chloronella* research for financial support."

"It's certainly not good."

I started the car and headed back down Sunnyside's driveway. "Besides the military, are there other sources for that kind of research?"

"The petroleum industry."

I pulled on to the road. "Petroleum? Why?"

"For one thing, they have a ton of money. Second quarter profits for the big five were almost twenty billion dollars this year."

I did a quick calculation. "My god. That's something like seven million dollars an hour."

"They need tax breaks, and funding local biofuel research is one way to do that."

"But don't renewable biofuels undermine their business? I mean, petroleum as a fuel?"

"Maybe. Maybe not. They could be planning for the future. Or they might want to control how the research is done."

That pricked something way back in my brain. Something important. I tried to dredge it up, but the idea stayed frustratingly out of reach.

"Does any particular company fund biofuel research?"

"Pacific Petroleum. They're big into it."

Pacific Petroleum. I stored that bit of information for safe-keeping.

Neither of us spoke for quite a while. Georgina's nasty words ran through my mind.

I broke the silence. "Want to talk about anything?"

"No. Not really."

Back at MOI, I pulled up behind Ted's truck.

Ted opened the passenger door and stepped out. He stuck his head back inside and tried to smile. "Thanks for taking me along."

"Happy for your company."

Ted reached over and squeezed my knee. "See ya later."

I drove home, absentmindedly rubbing my knee.

17

HARVEY BURST INTO MY OFFICE the next morning. She waved the *Spruce Harbor Gazette* above her head like a flag.

"You gotta see this."

"What on earth?"

She spread out page four on my desk and pointed to a photograph of Connor lifting Frank off the ground. It was hard to tell whether Frank was terrified or just surprised. But Connor was clearly livid. The title read "Local Citizen Saves Marine Biologist."The article described me as "a global warming scientist" who was attacked by "an unbalanced man who claims he's solved that problem."

"Where did they get this stuff about global warming?" I asked.

"That's what Frank was mumbling about, and the reporter heard him."

"Jeez, this is embarrassing."

My office phone rang. I spoke to the caller, ended with a time we could talk, and hung up the phone.

"Who was that?"

"Someone from the Portland newspaper. She wants to interview me, although I don't know why."

Harvey rolled her eyes. "Think about it, Mara. The story has everything. A clever scientist who does biofuel research goes nuts. And there's you—the equally clever female ocean-ographer who studies global warming and is attacked by the crazy scientist."

"Damn, Harv. You make this sound like a cross between a soap opera and a cheap mystery novel."

"Don't you get it? The press will *love* this story. I bet you'll hear from half a dozen newspapers."

She was right. I ended up doing phone interviews with the *Kennebec Journal* and a few other Maine papers, grateful for the opportunity to voice my views about climate change and global warming.

But I flatly refused a whole category of other requests. Speaking to any citizens' group was absolutely out of the question.

Not a soul knew this—not even Harvey. The fear of public speaking incapacitates me. I'm also utterly self-conscious about the whole thing.

I'm terrified to speak to a group of non-scientists I don't know.

At the start of my teaching career, I'd prepare like crazy and get sick to my stomach right before each lecture. Thank god that's behind me, and now I love teaching. And I'm okay at scientific meetings because I practice my talk over and over and know the type of questions peers will ask. In that way, I'm in control. The meeting with Gordy's buddies was uncomfortable for sure, but again I anticipated and prepared for the questions about climate change. And Gordy was there to help.

But talking to a bunch of total strangers is different. Put me in front of the public, and I get the shakes, my heart races, my voice gets squeaky. Glossophobia, the fear of public speaking, is my deepest, darkest terror.

And believe me, the irony of being Bridget and Carlos Tusconi's daughter—they were so relaxed and charismatic in front of any audience—has not escaped me.

So, no, I didn't talk to Rotary or any other group like them.

The MOI newsletter had concurrently featured the *Spruce Harbor Gazette* story to highlight scientists' struggles with climate change denial. Some of the grad students offered a "way to go!" in the hallway.

Predictably, one person was not happy about all of this. Seymour called me down to his office. As soon as I walked in the door, he held up the MOI newsletter.

"You're working hard to get publicity for yourself."

"What?"

"You should do research and write papers, not waste time on crap like this."

I was incredulous. "The guy from the *Spruce Harbor Gazette* just happened to be at the restaurant. What am I supposed to do if a newspaper calls me? Not talk to them? Wouldn't you worry what they'd think about MOI scientists if I did that?"

He grunted. I stormed out and nearly careened into Betty Buttz in the hallway.

"Seymour giving you a hard time?"

I growled, "You got that right."

"Bet you need some fresh air. Come on."

We went out the front door. She stopped and put a hand on my parents' plaque. "I knew them pretty well, you know. Back when MOI was established."

Frustration with Seymour evaporated. "You were friends?"

"Sure. Bridget and Carlos. With the fallout from that *Gazette* article, I bet you're following in their footsteps. Your mom, in particular, gave powerful speeches about marine conservation."

"Um, I don't give talks. You know, to the public. I'm—ah—not ready."

Betty pressed her lips together, and her bushy raised eyebrows compressed the wrinkles on her forehead.

She pointed to granite steps leading up to the door. "Let's sit for a while, over there in the corner."

I sat cross-legged. Betty straightened her stout legs out in front of her, and looked straight ahead. "Sometimes it's the hardest things we have to keep at. That's how you get strong."

I studied her face and waited for more.

"When I was in college, there were very few female oceanographers. In the U.S., women weren't even allowed on overnight oceanographic trips in the years following World War Two. Bad luck, you know."

"So how'd you do your research?"

"The taboo was broken in 1963, when a visiting Russian scientist on a research cruise turned out to be a woman. They couldn't very well tell her to go back home, could they?"

Fascinated, I said, "What happened?"

"The trip was a huge success."

"And then?"

"The taboo was broken, but for years some men tried the old excuses. I lobbied hard and won and sailed on many a month-long cruise in my early years and after that."

Betty turned toward me, gray eyes bright in her wind-weathered face.

"You're Bridget Tusconi's daughter. Wait too long, it might be too late."

With a grunt, Betty pushed herself to her feet and limped away toward the parking lot. I watched the square silhouette until she disappeared.

The painful task of reading and re-reading the draft of the grant proposal awaited me. I plopped down in my office chair, flipped on the monitor, and tried to focus on the logic of the proposal. My mind wandered. With a sigh, I closed the file and turned toward the window. The day was bright, but the most distant islands were hard to make out in an offshore fog.

Betty's decree troubled me. She was right. I'd used every manner of excuse to avoid public speaking. I was too young, too busy, not ready. All that was nonsense.

I was afraid.

Okay, but afraid of what? People who might criticize me? Questions I couldn't answer? I just didn't know.

I leaned back and shut my eyes. Betty's warning and my guilt had brought on a cold sweat. It took several minutes to get back to normal. I vowed to confront my public speaking problem very soon.

But first, there was a more immediate problem—my infuriatingly slow progress looking into Peter's death. What I'd found out was a jumble. Aloud, I said, "What do I really want to know?"

When I spoke with Connor and Angelo, it was Connor who'd voiced the key question. Were Frank's deception, Peter's suspicion, and what happened on *Intrepid* linked somehow? I'd told Connor the connection was impossible to imagine. I still couldn't see what it could be.

There *was* one person who might be able to help me. I knocked on Ted's office door.

He called out, "Come on in. It's unlocked."

I pushed the door partly open and stuck my head in. "Got time to talk?"

Ted turned, rubbed his eyes, and grinned. "Sure. I'd love an excuse to put off reviewing this paper."

"It involves Frank."

The grin vanished.

"Why don't you meet me outside on the pier? I've got some homemade cookies in my office."

Fifteen minutes later, oatmeal cookie crumbs lay between us on a granite bench on the dock. I'd told Ted about MOI's next speaker.

"The grad students invited Stan Huntley for Monday's lunch talk. He's great. We've known each other since our MIT days."

"Good. But you didn't invite me down here to talk about that. What's up?"

"Like I said, it's Frank."

Ted looked out at the harbor. "What do you want to know?"

"Well, where is he?"

"In psychiatric rehab close to his parents in Wiscasset."

"Have you visited him?"

"Not yet. I'm thinking of going today."

"Ted, I'd like to go with you. To talk with Frank."

He turned toward me. "What? Why?"

I touched his arm. "Please. I want to know why he did it. Why he tried to pull off that hoax."

He frowned and knit his brow. "What's so pressing?"

"Maybe there's a connection with Peter."

"Mara, I have to think about Frank. Seeing you so soon wouldn't be good for him."

I pulled my hand away. "Just listen. What if he's not alone in the *Chloronella* fraud? What if someone pressured him into doing it? Someone like Georgina."

"But why—"

"I'm not sure yet. But if that's what happened, I won't file a complaint against him."

Ted got up and walked to the edge of the dock, hands in his jean pockets. "No felony on his record—that'd be huge."

I looked up at Ted expectantly.

He ran his fingers through his hair. "Okay. Maybe we can visit this afternoon."

I climbed into Ted's truck. Neither of us was in the mood to talk, so the hour-plus ride down to Wiscasset gave me time to think. The last time I saw Frank, I was in shock on the asphalt of a parking lot. This encounter was going to be stressful. I'd feel angry. Maybe even scared. Or both.

The mental health facility wasn't at all what I expected. Isolated, with a view of Wiscasset harbor, its red brick exterior had many windows, and an expansive porch dotted with chairs looked inviting. Clearly, Frank's parents could afford the best for their son.

Ted signed us in at the front desk and we followed an attendant to a garden area off the dining room. He motioned to a man on a wooden bench, adding that he'd be back in a half-hour.

Ted sat down on the bench opposite his childhood friend. I chose a wicker chair to the side. In a white T-shirt, khaki pants, and white tennis shoes, Frank could be in Central Park on a Sunday afternoon. He took a sip from the mug in his hand.

Ted said, "You look good. You okay here?"

"For this kind of place, it's not so bad."

Ted asked about Frank's parents. I studied Frank as he answered. It was impossible to relate this ordinary-looking guy chatting about his mom and dad to my attacker. I didn't feel angry or scared. Instead, I sensed something off about the man.

Ted said, "Frank, this is Mara Tusconi."

I didn't know whether Frank had noticed me. He swiveled his head in my direction.

I put my hand on the bench and leaned toward him. "Frank, do you know who I am?"

"Sure. Lady with the lobster."

I swallowed the anger boiling up and decided to jump right in. "Frank, Ted and I know you've been adding fertilizer to the *Chloronella* tanks. We've got isotope data showing that. We'd like to know why you did it—why you fertilized those tanks and let on you'd bioengineered the algae. And if someone pressured you into it."

If Frank was curious about how we got the water samples, he didn't show it. He narrowed his eyes. "And if someone did?"

"Well, I might drop charges against you if you'd explain why a researcher on a biofuel project would undermine the work like that."

Frank's neck flushed red, lurid against his white shirt. He spat out his reply. "Nobody tells me what to do."

Ted sucked in his breath.

Frank kept going, his speech rapid fire. "Christ, you're so naïve. You think it's about the research? I just regret I got caught. More time in this country club. Then I'll live the high life."

Ted tried to reason with his friend. "But Frank, you might go to prison."

"Yeah. But I have no record. I'll be a good boy. Get out, live like a king."

I told Ted he'd find me in the lobby, and left. We were a good five miles out of Wiscasset before I said anything. "Do you want to talk?"

"I keep thinking what a great kid he was. The guy back there is a stranger."

"What's he saying? Living the high life when he gets out of prison?"

"Who knows? Either he really is crazy or somebody paid him a lot of money to scuttle algal biofuel research. And he'll get more later."

"But who'd do that? Pay a scientist to make everyone think he'd engineered a super-seaweed? Since he didn't, that would come out. It doesn't make any sense."

"Someone with a whole lot of money, apparently."

Ted flipped on the radio and tuned into the local public station. We didn't speak for the rest of the trip. Ted listened to the news while I gazed out my window and pondered Frank's claim. I wasn't sure the trip moved my investigation of Peter's killer along.

But at least I knew who Homer's assailant was.

18

I WALKED INTO MY KITCHEN. The light flashed on my answering machine. Stan Huntley wanted me to call him back via video chat.

Stan Huntley, a well-known physical oceanographer from MIT, was a dear friend. In the final year of his PhD research, I started mine. Stan helped me through some tough times. My experience as a woman in the world of science was nothing compared to Betty's, but it was bad enough. Over coffee or beers, Stan patiently listened to me gripe about men who hinted that females didn't have the smarts or stamina to succeed as oceanographers. I'd trusted him with my most secret worries and would do anything for him now. When I was awarded my first big grant, he was the first one to call with congratulations. Once a year or so we had dinner at meetings we both attended.

Minutes later, we were face-to-face across cyberspace. With hollow cheeks and smudges under his eyes, Stan looked dreadful.

"Mara, I have to cancel my visit."

"I'm so sorry. Are you sick?"

"No, not like you think."

"Stan, what's the matter?"

"Climategate all over again, Mara. It'll be in the papers. I'm one of a dozen scientists whose email's just been hacked."

"For the second time."

"Yeah. I was on the Prospect Institute's list. This time it's an absolute nightmare. MIT's doing what it can, but this will be a mess. A deluge of Freedom of Information requests, lawyers. Appearing before an investigation panel at NSF and on the Hill."

"But why you if there's eleven others?"

He leaned toward the camera. "This time I'm the main target for these fascists. I used the word 'trick' in an email—you know, as in a novel way to solve a stubborn modeling problem. You can imagine what the climate change denial PR folks've done with that."

I groaned. "You're a trickster. Scientists are tricking the public. It's all a trick."

"Right. And I've got to stand before this congressional panel with people who deny evolution, never mind climate change, and try to explain that mathematicians use the term 'trick' all the time to explain a clever solution."

"And the irony is that *your* email's been hacked by a thief. You've done absolutely nothing wrong."

"Yeah."

"Who's behind the hacking?"

"A front called Freedom America."

"Who backs them?"

He waved his hand. "The usual. A couple of conservative, extremely wealthy people. And some petroleum companies."

"Petroleum? Any companies in particular?"

Stan dropped his head and massaged his neck. He spat out his answer. "Pacific Petroleum."

"How do you know?"

"You can use IRS records to identify key supporters of climate change denial groups. Last year alone, Pacific Petroleum gave fifty million dollars."

The hairs on the back of my neck brushed against my turtleneck.

"Stan, I'm so sorry. Wish I could do something for you."

He shook his head. "Stay away from companies like Pacific Petroleum, Mara. That's what you can do for me."

I woke up early the next morning, took a shower, and toweled off. Stan's warning worried me. A mug of coffee beside me on the kitchen table, I entered "Pacific Petroleum" into my browser and skimmed the first entries.

I called Harvey at home. She picked up on the first ring.

"It's Mara. Do you have time for me to come by for a bit?"

"Sure, but—"

"Be at your house in ten."

I pulled into Harvey's driveway and bustled into the kitchen. Harvey leaned over her counter, a cup of coffee and newspaper in front of her. She raised an eyebrow.

I climbed onto the stool next to her. "I talked to Stan last night. He has to cancel."

"Good morning to you too."

"Listen, it's important." I told her about the latest harassment he had to deal with. "According to Stan, a company called Pacific Petroleum is behind some nasty climate change denial schemes."

"That's just dreadful. God, the poor guy."

"He looks terrible."

"And you're here," she glanced at the kitchen clock, "at six-thirty because—?"

"I'm wondering if that company's in the middle of our mess."

"What?"

"This'll sound crazy, but just go with me for a minute. Someone connected to oil interests could've paid Frank a ton of money to create an elaborate deception. You know, to make people think he'd engineered a super-seaweed when he never did. Think about it. It would completely undermine a high-profile biofuel project. Shake people's faith in alternative fuels—that's a cunning strategy."

"You think it might be this Pacific Petroleum?"

"It's possible. Get out your laptop."

I leaned over her shoulder as she sat at her kitchen table and recited facts about Pacific Petroleum. With annual revenues topping $400 billion, the company was a huge oil and gas conglomerate. Their website boasted a high profile, multi-million dollar program called "Plus Petroleum" that funded projects focusing on wind, tidal, and biofuel, including algae.

"Damn," I said, taking in the fancy calligraphy and intertwining green leaves on the Plus Petroleum page. "Cute look."

Harvey pushed back from the computer. "But you can't mean Pacific Petroleum actually awarded Frank a legitimate grant."

My hand was already on the doorknob. "They would've paid him secretly. Hey, thanks for the help. Think I'll pass this by Connor."

Pink colored Harvey's cheeks. "Connor? Why Connor?"

"He used to be a cop. He'll tell me if this idea is totally nuts."

I called Connor on my cell as I walked down Harvey's driveway. He was about to take off to go fishing.

"I'm at the town dock. Take a quick spin with me."

A half-hour later, I stood next to Connor at the center console of his boat. He'd given me his sou'wester, but the early morning air chilled me anyway. Just outside the harbor, Connor upped the throttle and the boat took off like a racehorse. We slapped against the waves, and I whooped when a shower of salt spray soaked my flying hair and ran down the sou'wester. Connor cut the motor, and we slid to a stop below Angelo's property on Seal Point.

He pointed up toward the house. "From up there, you forget how unforgiving Maine's coast is."

I followed his gaze. The granite cliff below Angelo's lawn was at least a hundred feet straight up. At the water's edge, waves crashed against boulders twice the size of my car.

Connor started the engine, backed away from danger, and slid the lever to neutral. Against the gurgling motor, he said, "So what's up?"

"I'd like your help with a nefarious idea."

He grinned. "Nefarious? You bet."

I went through the whole story—from the conversation with Frank to my suspicion about Pacific Petroleum's biofuel scheme at Sunnyside.

Connor pushed wet curls off his forehead and pulled on a baseball cap. "Problem is you have no proof. Your idea seems possible, I suppose. But it's a big leap and without proof, it might as well be mischievous wee folk."

I peered into cloudy water where the swirling current had stirred up the bottom. My enthusiasm faded. "I get what you mean."

"Besides this Frank, is there someone at the seaweed place you think might be involved?"

"Georgina Hamilton, John's wife."

"What's she like?"

"Smart, city lawyer type—who keeps her husband right under her thumb."

"So you're thinking she might be in on this with Frank? Be the contact with the oil company?"

I ran my hand across the salt-crusted console. "She's on oil and gas corporate boards. Maybe someone from Pacific Petroleum approached her."

A tern swooped into the water and came up with a flapping fish. On a nearby boulder a gull let out a hoarse laugh. As we bobbed out there in the open ocean, the whole Pacific Petroleum idea seemed a lot less likely.

Connor brought me back. "And how does this connect with what happened on the ship?"

"Big oil fronts climate change doubters. The scientists on the *Intrepid* cruise work on climate change and global warming. Maybe somebody got paid a lot of money to sabotage that trip."

"A crewman? I know those guys and can't imagine—" He frowned.

"Did you just think of someone?"

He rubbed his neck. "What? Nah. Didn't you say there was a visitor on the ship? John somebody?"

"John Hamilton, Georgina's husband. No, not him. He's a good guy. Kind of a milk-toast, actually. And he's totally dedicated to alternative fuels. If Georgina is behind this, she's doing it on her own."

"Why would this Pacific Petroleum go after MOI?"

"MOI's known for climate change research. The institution was featured on a big deal TV show a while back. Come to think of it, Peter was interviewed. He talked about the Greenland ice sheet—that if it melted sea level could rise twenty feet. Can't believe I didn't think of that before."

"Lots of maybes, Mara." Connor looked at his watch. "We've been out here nearly an hour. Time to get you back."

I sighed. "This Pacific Petroleum idea. Where can I go with no evidence?"

Connor slid the lever on his throttle forward and we headed toward the harbor. He tugged his baseball cap down. "I'd say you're stuck."

19

CONNOR DROPPED ME AT THE town pier and sped off with a wave of his hand. He was right. I had to be honest with myself. My speculations about Georgina and Pacific Petroleum were just that. Even if I were dead on, there was no way to prove it.

I pondered that problem on my way to the biology building. I stopped in the office and told Seymour's assistant about Stan's cancellation. Up in my office, my desk was cluttered with a six-inch pile of scientific papers waiting to be read, two to review, and a student's PhD research proposal. Besides that, segments of the new grant proposal had to be woven together. Pacific Petroleum's possible dirty tricks would have to wait.

Later that morning, my new grad student, Alise, knocked on my door and stuck her head in. A welcome distraction, the redhead radiated excitement.

She fell into the seat next to my desk. "Dr. Tusconi, I'm thrilled to be here at MOI, especially as *your* grad student."

The joy was infectious, and we grinned at each other. With Graduate Record Exam scores in the top one percentile, she was real, real smart—although some scientists might never guess to look at her. Today, her first official one at MOI, she was conservatively dressed in jeans and a long sleeved white shirt. But her college mentor, a close friend, informed me a while back that Alise typically wore T-shirts with slogans like *Take Care of Mother Earth: We Don't Have a Backup*. Besides that, she sported fish tattoos on her upper arms and listened to loud hip-hop music when she sorted samples in the lab. While none of this was particularly unusual for people her age, Alise would stand out in a group of MOI grad students.

I didn't care. To me, she looked like a young woman with the intelligence, curiosity, drive, and spirit to make it through the grueling five plus years before her. And besides, watching Seymour try to figure her out was going to be fun.

"Can't believe I'm already going on a research cruise," Alise said. "And one that focuses on marine acidification."

"Speaking of which, what did you think about that paper I sent you about acidification and oyster larvae?"

"Well, it's the first credible study that shows larvae might be able to adapt to low pH and grow normal shells. Nice to have a bit of positive news."

Having Alise as my student was going to be terrific. "Come on. The chem labs are on the second floor. You need to meet Harvey—Dr. Allison—and other students going on the cruise."

After I'd confessed my worries to Harvey about supporting a new grad student, we hit on the perfect solution to the problem. Her research assistant had moved to California, so Harvey had money to pay a student to run chemical analyses for a year. Alise had aced physical chemistry, and her thesis integrated biology and chemistry in a study of ocean acidification. Alise served Harvey's needs perfectly.

The second floor bustled with scientists and technicians preparing for the next day's departure. Plastic carryalls, cardboard boxes, and trolleys loaded with computers and scientific equipment lined the hallway. Since the cruise focused on chemistry, not biology, I wouldn't be on board. Harvey would take excellent care of Alise, and it would be the perfect opportunity for them to bond.

Alise and I walked down to the elevator where Harvey and a student struggled with an overloaded cart. The four of us managed to get the cart into the elevator car without breaking anything.

Harvey pushed the down button and turned toward Alise. "So pleased you'll be on this cruise. We'll have plenty of time to talk. Come on, let me introduce you to the other students."

The pair walked down the corridor, heads bobbing enthusiastically. On my way back to the stairs, I passed the computer lab and looked in. Empty. I guessed everyone was busy preparing for the cruise. Today I'd let myself look at the buoy temperature data. What the hell. Maybe a different computer would be good luck.

I logged on, called up data for one of the buoys we deployed, selected "specialized plots," and closed my eyes for a moment. Heart pounding, I opened them and fixed on the uppermost plot—temperature. The day-by-day graphs were a series of saw-tooths, each peak representing one day as temperature rose with the sun and dropped again at dusk. Different colors represented different depths—red for the surface, orange below that, and so on. I focused on the midday temperature peak and used the eraser end of a pencil to trace those values from day to day on the monitor screen. My eyes widened, then narrowed as I leaned in closer to the monitor.

Temperature increased *a lot* each day—especially at and near the surface.

I'd have to check, but the rate of daily increase looked like it might even exceed last spring's record. In quick succession, I called up data for the other buoys. The data were similar.

I powered off the computer and paced the empty room. This was good and bad. Evidence of record-breaking warming once more in Maine's waters would make the NSF proposal with Gordy a lot stronger. We'd have evidence of a trend, not just a single year event. That was the good part. But, as Gordy pointed out, a hotter ocean was bad news for fishermen, fish, and a whole lot else. I had to keep that sobering thought in mind.

I reached the landing on the third floor. Ted stood in front of my office door, about to knock.

I called out, "Be right there."

Ted backed away from the door and leaned against the wall, hands in his jean pockets. He looked like a model—jeans for the mature male. I grinned.

Ted said, "What's so funny?"

"Ah, water temperature's on the rise, just like last year."

"I want to hear all about it. Hey, do you have time for a break, Mara? I'm pretty much all set for the cruise, and there's something I want to talk with you about."

We sat at one end of the MOI pier. Once more, Ted leaned against a piling, his legs outstretched. I sat cross-legged with a direct view of *Intrepid*. People walked up and down the gangway in preparation for the cruise.

"You leaving at dawn?"

"Yeah. Wish you were going?"

I fixed on the spot where the winch crushed Peter. "It's a chemistry cruise, but I'll never be on that ship without thinking about Peter."

"Me too." Ted threw a stone into the water. The ripples widened and disappeared.

He asked, "So what's up with the water temperature?"

"The rate of increase is accelerating. A little later than last spring."

"Great the buoys are so useful. Look Mara, there's something I want to tell you about. You can't discuss it with anyone. Not even Harvey."

I raised an eyebrow but said nothing.

"Last week I drove to Boston to secretly meet with a group of young climate change researchers. We're disgusted with the harassment and personal attacks. Stan's just the latest. People's lives are being threatened. It's completely out of hand."

"I agree. How'd it go?"

"The meeting was very small—there are ten of us—and arranged by word of mouth. No email to hack. We met in a hotel. Nobody knew what we were talking about."

"Which was what?"

"This intimidation *has* to stop. We spent two days figuring out how to do that. We have to expose these people—and groups—whoever they are. For one thing, they're abusing Freedom of Information laws. This isn't about sharing research data. It's about clogging up the works of climate change research and driving scientists out of the field."

"Sure. But what can we do about it?"

"We need the press on our side. They've been trying too hard to be neutral—showing both sides. But the public needs to understand thousands of scientists are persuaded that the climate is changing, with humans largely to blame. A tiny minority disagree, and many of them have questionable motives."

"And you're going to do that how?"

"We need someone the press will love—a scientist who does research on global warming, is smart and articulate, would appeal to the general public, and who is so obviously honest any harassment would be exposed for what it is."

"Have you all identified somebody?"

"Yes."

I waited.

"You."

I laughed. "Very funny, Ted. You can't tell me who it really is?"

"Mara, you're the perfect person."

This time I didn't laugh. "Surely you're joking! I'm not famous. I don't have clout. I have no experience talking to the press. This is serious stuff, and you have to find the right person."

"Mara, you *are* the right person. You've just been in newspapers all over Maine because you care to your core about integrity in science. You're a bright young scientist, and you do climate change research."

My hand went to my heart. He was deadly serious.

"But what about you?"

"I haven't been in the news, and I'm not great talking to the public. You are. You teach oceanography to five hundred undergraduates at a time. You're really good at it. Some come into the class hating science, and you turn them around. They say your course was the best one they took. I can't think of a single science professor who can claim that."

It was true. I even got cards from students thanking me for the class, excited that they could figure out science in the news. Many mentioned the climate change lectures in particular. But what Ted proposed was totally different. He was asking me to do the one thing I avoid at all costs. I had to get out of this. I could admit my fear of public speaking, but my anxiety was mortifying.

I desperately tried to think of an excuse.

"One big problem. Seymour's already on my case about publicity. He'd tell me I have no business doing this."

Ted snorted. "Seymour is a jerk. He's just jealous, Mara. Nobody cares about his research on squid. It's old stuff, and he knows it. The irony that he's got your parents' professorship can't be lost on him."

No other excuses came to mind, so I agreed to listen. Ted described "the plan"—a high profile session with several speakers, including me ("or whoever," I added), for the press. It would be part of the National Association of Ocean Scientists—NAOS—meeting at MOI in less than a week.

"But I haven't submitted an abstract to give a talk. That would've been due months ago."

"No problem. You'd be invited as a lead speaker."

"Ted, I don't have experience dealing with the press! This isn't a small deal."

"Several researchers in the group have been through the National Association training. They feel confident you could get up to speed in a day."

This was looking very bad. I tacked again. "So after your spokesperson gives a talk and gets some press, then what?"

"We're hoping that some journalists will have the integrity and clout to step up to the plate and show what's really going on here. Remember Woodward and Bernstein?"

"Of course. They exposed the Watergate scandal for the *Washington Post*."

"Well, we want to find their present day reincarnation. What do you think? Is this something you'll at least consider?"

Stan had warned me about Pacific Petroleum. "Ted, I'd be a sitting duck for the bad guys."

"We talked a lot about that, of course. We're working with excellent lawyers who know a lot about Freedom of Information law, for one thing. They'll do it *pro bono*."

Ted's pleading eyes just about melted my heart.

"Ted, this is huge, just huge."

"Please, think about it. We can talk when I get back." Ted got up. I watched until he disappeared around the corner of our building.

I swung my legs over the edge of the pier again and threw pebbles into the water, watching them disappear from sight.

I was rinsing dinner dishes that evening when someone knocked on my kitchen door.

I pulled the door open. "This is a surprise."

Ted's grim demeanor said it all. "Can I come in? There's bad news. I want you to hear it from me instead of over email or something."

My hands shook as I shut the door behind him. "What's happened? Is it somebody from MOI?"

"No. It's Stan Huntley."

My hand went to my mouth.

"He's in hiding."

20

I FELL INTO A CHAIR next to the kitchen table. "Hiding?"

Ted took the opposite chair. "He got threatening emails again from some climate change doubters group. Vicious ones. Some threatened his kids. MIT contacted the FBI. That's all I know."

"I can't believe this. How do you know?"

"That group I told you about this afternoon. One of the guys—Stan's colleague at MIT—called us. Since you're good friends with Stan, I wanted to tell you in person."

"I appreciate that...I'm stunned. How can people get away with this garbage? Threaten scientists' *children*?"

"I know. I know."

After Ted left, I wandered around the house and ended up on the living room couch. Stan was bitter about the harassment, but who'd ever think he'd need FBI protection? At midnight I went to bed, but sleep wouldn't come. Finally, I rifled through the medicine cabinet for a sleeping aid and took it with a slug of water.

The next morning I woke with a start. The clock said seven. The grant proposal deadline was looming, and I'd overslept.

I threw off the cover and fell back on to my pillows. It was frustrating and sad that I couldn't contact Stan, but he would be the last person to stand in the way of my research. I'd send him a special prayer in my own way but now had to pull myself together. A colder than normal shower and two cups of coffee got me going.

The core of the proposal was still in draft form. My task was to pull it together so Gordy could see it. I also needed to recruit the right scientists for the project. They had to do climate change research and be interested—no, psyched—to work on such an unusual venture.

I revisited the list of scientists I'd listed on my whiteboard and added a couple more. For the proposal, we needed a commitment from six.

My email message to the scientists described the NSF program and our plan. We wanted to bring two disparate groups together who'd "mutually develop strategies so Maine's fishers could modify fishing practices using climate change research." I sent the message into cyberspace and worked on the proposal draft. Two responses came in. Neither encouraging.

Dr. Tusconi: While I admire your creativity, I don't have much confidence in a project like this.

Dr. Tusconi: Sorry, too busy for a venture with such an uncertain outcome.

"Too early to be discouraged," I told myself. But, of course, I was.

As it turned out, much worse emails were on the horizon.

I was lacing up my running shoes for a brain-clearing jog when the first message from Harvey registered an alert on my email program. I jiggled the mouse to wake the computer and read:

Want you to know about some bizarre accidents on the ship. Early this morning a plastic carboy filled with distilled water crashed down from the top shelf in the lab. Luckily, nobody got hurt. Of course, we'd secured it.

My mouth went dry. On a ship, anything not held in place could end up on the deck in short order. Harvey would've spotted an unsecured carboy in a second. I yanked my desk chair out of the way and leaned close to the monitor to read.

The next incident could've been a lot worse. Alise walked into the lab and found a bottle of acid sitting out on the lab

bench. We were absolutely sure it'd been safely stored in the
chemical cabinet. The bottle could've smashed on the deck.

I imagined the scene. A broken bottle of acid would've
filled the lab with poisonous fumes. Worse, whoever came
in to check on the smell might slide on the liquid and fall
into a puddle of acid. My god. That could've been Alise. My
hand, still on the mouse, shook.

I read the rest of the email.

The captain is extremely concerned, and we're all on edge.
Keep your fingers crossed. Harvey

I closed my eyes and willed everyone aboard *Intrepid* a
safe trip. On my way down the stairs, I had to steady myself
with the railing.

I ran along the Spruce Harbor waterfront and followed
the hilly road tracing the shoreline. At the crest, a dirt path
took me down to the rocky headland opposite Seal Point
where Angelo lived. Wing Point, basically a long granite
cliff, was too steep for houses, so it remained a preserve for
hikers and birdwatchers. I reached the end of the path and
stopped to catch my breath.

By that point, my alarm had morphed into outrage.
I shook my hands and arms to get rid of accumulated
adrenaline.

Once more, somebody on *Intrepid* tried to sabotage a
research cruise. Who on that trip had been on the previous
one? The list was long. Harvey and Ted, Ryan as first mate,
plus the rest of the regular crew and a few of the grad students. I wasn't sure about Cyril White, the photographer. A
couple of people weren't on board now—me, Seymour, John
Hamilton. Of course, that didn't mean much. Someone like
Georgina Hamilton could pay one of the crew to sabotage
the trip.

The big question was why.

Looking to the east, I imagined *Intrepid* bobbing in the
waves out there somewhere and wondered how Harvey and

Alise were doing. Naturally, they'd be nervous. I winged them a prayer. You don't think about it when you're busy working, but a ship at sea is a tiny, vulnerable thing. Much can go wrong, especially if someone on board makes sure it does.

In my scientific circle, things were very, very wrong. No one on a research cruise should be in danger—not Peter, not Harvey or Alise. No one should need FBI protection because they study climate change.

And I'd been offered an opportunity to do something— help with a high profile meeting for the press. Betty was right. If I waited too long it was going to be too late.

Goosebumps rose up on my arms. The wind had shifted and that—plus the incoming tide—made for an unsettled sea. Hundreds of feet below, waves crashed against the granite ledge, sending spray thirty feet into the air. Over and over, the sea pulled back and smashed against the rocks. The show mesmerized me until dark clouds covered the sun and I shivered in my running shorts and T-shirt.

It was time to leave, but I took a moment for Stan first. Since geology was his hobby, I searched around for a classy rock specimen and settled on a chunk of bright pink granite dotted with crystals of black hornblende.

I held the rock in the palm of my hand and said, "Stan, pink is the color of hope and black the evil that hounds you. Both of us must believe in hope."

Below me, the sea pulled away in preparation for the next crashing wave. I threw Stan's rock into the mélange just as the surge began again.

By the time Wing Point was behind me, there was no question what I was going to do.

Back in my office, I opened up my email program. The message, disguised to foil hackers, was short:

Ted—the day after tomorrow looks perfect for kayaking. Are you free? We have lots to talk about. Mara

Ted replied right away:

Terrific. I look forward to kayaking with you.

The *Intrepid* cruise was only an overnight trip. The ship pulled up to the MOI dock late in the afternoon. I was on the pier to meet her. Alise spotted me, waved from the deck, and was the first one down the gangway. She'd already put the acid bottle incident behind her.

"Dr. Tusconi, it was a fantastic trip. I learned so much. And how far you can see—it's astonishing."

I boarded the ship and found Harvey taping boxes in the lab. She gave me a tired smile.

"How'd the rest of the cruise go?" I said.

"No more problems, thank goodness. I'm done here. We can talk in my lab."

"Is Ted around?"

"I believe he's on the forward deck."

"Meet you in ten minutes."

I found Ted and helped him secure a bungee cord around an oversized tote. When we'd finished he stood, arched his back, and rolled his shoulders.

"Long night?"

"Yeah, but a productive one." He shook his head. "Harvey told you what happened. Thank God it wasn't worse."

"About the kayak invitation for tomorrow. We can talk out there." I pointed toward the bay. "But if you're tired—"

His smile lit up his face. "I'll be fine in the morning. That sounds great."

We arranged to meet at my house at nine. I followed a slow-moving cart down the gangway and hoofed it up to Harvey's lab. She closed the door and went back to organizing boxes and equipment from the trip.

I asked, "Any idea who did it?"

"Three, four, five. Good." She shrugged. "After the acid bottle incident, I sat in the ship's mess and tried to figure out who might be guilty."

"And?"

"That bruiser of a guy you complained about—the one with liver-colored eyes. He was on the ship again. He's friendly with a couple of other rough-looking characters. But you can't tell much by that."

"His name's Jake."

"The photographer was on board as well."

"Cyril White." I bit my lip. "Odd he'd go again. He claimed he was so busy. Damn it, Harvey. Whoever's responsible could've slipped into the lab, done the deed, and slipped out again in less than a minute. Maybe when you were in the mess. It could be anyone."

Harvey leaned back against the lab bench and rubbed her neck. "Not knowing is extremely frustrating."

"Ryan was on board?"

"He was. Guess they concluded the buoy was just an accident. But he didn't operate the winch. And Mara, he's a walking ghost."

"He feels guilty as hell."

I followed Harvey out of the lab and stood in the hallway while she dug around in her purse for her office key.

"I'm kayaking with Ted tomorrow."

She dropped her key and picked it up. With a toss of the head, she said, "Really? Why?"

"Fill you in, promise."

Ted's truck rolled to a stop in my driveway at 8:59 the next morning. It took three trips to get two kayaks and the rest of the gear down to the beach. Ted used my "guest" boat, a red fifteen-footer. Gentle waves made for an easy launch.

"Where're we heading?" he asked.

I pointed to an island a mile out. "That's Cove Island. Straight shot about one-hundred-sixty degrees, but it's safer if we hug other islands along the way. High tide's midday, so we can paddle back into the cove. It's pretty—beds of eelgrass, lots of starfish, mussels. Usually eagles nesting. We can have lunch there."

"It's a MITA island?"

"Yeah," I said. "You a member?"

"Ah—no."

"Get with the program, Ted. The Maine Island Trail Association protects hundreds of Maine islands. Sign up when you get home, or I'll come after you."

Ted threw back his head and laughed. "I'm sure you will, Mara."

On the way out, we relished being in our two little boats atop the ocean. The water sparkled as we slid through a calm sea with a light following wind. We talked some but mostly enjoyed the silence—the only sound the swish-swish of our paddles through the water. At one point, a harbor seal popped up out of the water right next to the boats and surprised the daylights out of both of us.

We reached the island. In the cove, eelgrass and kelp swayed inches beneath the boats in crystal-clear water. We pulled the kayaks up to sand still exposed at high tide. Soon, our gear lay scattered around us as we leaned back against an old piling and ate turkey and Swiss cheese sandwiches.

"Man, this is heaven," Ted said.

Eyes closed, I'd been thinking the same thing. "Mmmm."

"So you've changed your mind?"

I sat up and turned to face him. "One of our colleagues is dead and another's hiding out somewhere. My grad student could've gotten badly hurt on that cruise. I still have serious doubts I'm the right person, but I have to do something."

I stood, brushed sand off my legs, picked up a handful of stones at the water's edge, and snapped my wrist hard to throw them, one by one, out into the water.

"This huge oil company conglomerate—Pacific Petroleum—" Snap. "—is pouring money into climate change doubters' crusades." Snap. "Their tactics are goddamn sickening. One scientist got a letter with a powder like anthrax." Snap. "There's even a billboard on which the Unabomber says global warming's real." I grabbed a rock the size of a bowling ball and hurled it into the cove.

Ted got up and stood beside me. He said quietly, "Mara, I've never heard you speak like this." He put his hand on my shoulder, and I turned toward him. Something shifted inside, and for an instant there was a roar like waves crashing on the beach. I stepped back.

Ted cleared his throat. "So why don't we start with your talk? Take it from there?"

"Right. Yes, let's."

"The meeting's title is *Global Warming: America's Leaders Tell It Straight*. There are two other speakers—guys from the military and firefighters' association."

Suddenly, my back was wet with a cold sweat. I swayed. Ted said, "You okay?"

"Just need to sit." I walked back to the piling and lowered myself onto it.

Ted continued. "I want you to meet some guys who were at that Boston get-together. I'll invite them to my house for dinner tonight. And you have to meet with the trainer."

I frowned. "Trainer?"

Ted opened the hatch cover on his kayak and pulled his phone out of a waterproof bag. "Give me a minute."

I waited.

"Just sent off the message about dinner. The trainer. I mentioned that to you. She helps scientists talk to the public."

"But Ted, the meeting's only a few days away."

"Mara, you're the most energetic person I've met. And brave." He swept his arm toward the ocean. "You're out here in your kayak in April and don't think a thing about it. You can do this."

My eyes scanned the cove, from the tips of the tallest spruce to seaweed-covered boulders. *I can do this, I can do this.*

We landed back on the beach where we'd launched. We carried the gear up to my house, and Ted checked his phone.

"Terrific. We're on for tonight." He headed for his car. "Mara, that was great," he said over his shoulder.

An hour was just enough time for me to shower, dress, and run over to Angelo's house to drop some dried chips for his outdoor smoker. I opened the kitchen door to the familiar smell of raw fish. Angelo's success on the water was evident in the sink.

"Hi, sweetheart. I'd give you a hug, but my hands smell like mackerel."

I watched my godfather cut the heads off two-dozen foot-long fish, stick his knife in the anus of each, rip open the underside, and pull out the guts. Each fish took thirty seconds. Smoked, they'd taste delicious.

As Angelo washed his hands, he said, "There was something on the VHF about *Intrepid*, but I couldn't make it out."

I leaned against the countertop and described what had happened on the ship and with Stan.

Angelo dried his hands on a dishtowel. "This is a bad, bad business, Mara."

"Ted was aboard. He invited me to speak at a meeting in a few days. It's about climate change. For the press."

"Really? This is the first I've heard about that."

"It's happened very fast." I hugged myself.

"I don't recall you doing anything like this before."

"Um, it's hard for me. I'm a heads-down person and would, you know, rather do my research and teach my classes."

I couldn't admit, even to my godfather, that I was scared to death about speaking to an auditorium full of strangers.

"But—?"

I met his eye. "But this is a defining moment for my generation of climate scientists. I can't walk away. Stan's the only hacked scientist who's in hiding as far as I know, but last month a researcher's inbox was flooded with hate mail directed at him and his wife. Can you imagine the paranoia? When I emailed Ted, I worried someone might read it. That felt awful."

Angelo's eyes widened. He threw the dishtowel on the countertop.

"Christ, Mara. I thought we were done with this. If these people are fanatics who don't care who they hurt and you give that talk…. *How* can we keep you safe?"

Taken aback by the intensity of Angelo's reaction, I said, "We'll work on that together. Promise."

21

AGAIN I LEFT ANGELO'S HOUSE feeling crummy because
he was worried, it was my fault, and I didn't know what
to do about it. I'd call him first thing in the morning.

But now I had to meet Ted's secret colleagues.

Ted lived ten miles inland from the coast. I'd never
been there, and the drive was a pretty one through spruce
and hardwood forest broken by villages and lakes. But as
my car got closer, my nervous stomach made itself known,
and it was harder to enjoy the view. To shake off tension, I
stopped and walked along the shoulder of the road. I didn't
know exactly what Ted's friends expected of me, and maybe
they'd be disappointed. I couldn't admit to my fear of public
speaking, so it felt like I was harboring a big secret.

I climbed back behind the wheel and soon spotted Ted's
red mailbox. His driveway led to a cottage on a rise sur-
rounded by open fields of grass. By the time my car slid to
a stop, a cold sweat had soaked my back. Walking toward
the house, I took in breaths and slowly released them like
we did in yoga class.

A classic old cape, Ted's house had shingles weathered
gray and two sets of windows on either side of a brick-red
front door. My hand was nearly on the knocker when Ted
opened the door. I stepped into a room with worn wooden
floors and cream-colored plaster walls. There was a large
central brick chimney with fireplaces in the kitchen on one
side and a living room on the other.

"Ted, this is charming."

"I know how lucky I am to live here. Come. We're comparing microbrews and would love your opinion."

He led me into the kitchen where two men stared at a row of little glasses lined up on the wooden counter. They straightened up when I entered the room.

"Mara, meet Tony Visconti and Mark Clements. Tony is from U. Maine, Orno, and Mark's from Boston University."

I shook each man's hand. They were a mismatched duo. Tony, a suave Italian with striking brown eyes and olive skin, obviously liked his drink and sported a belly that showed under his shirt. Mark, the slim, loose-jointed athlete, probably biked a hundred miles a day. The comical contrast helped me relax a bit.

Tony motioned me to join them. "We've got ten microbrews here from all over the Northeast. We're comparing hues. The dark ones are stouts and light ones lagers."

I squatted down. "I had no idea beer could be so dark."

Stirring chili at the stove, Ted caught my eye and winked. I tasted an especially sour Belgium beer—and Tony and Mark watched me intently. They laughed as I puckered my lips. I was already part of the group.

We sat around the old pine kitchen table and ate chili and cornbread. The room felt familiar to a New England girl like me—pine floorboards, plaster walls, Wyeth prints, the wood stove. You could sense the generations past who loved this home.

I felt more at ease. These were smart, funny, decent men, and we had much in common. Tony and Mark also worked hard and loved what they did.

The symposium didn't come up until we'd eaten ice cream and strawberries. Mark explained they'd invited some real-world speakers.

"Who, exactly, is your audience?" I asked.

"Good question," Tony said. "The press and general public. Since the sponsoring association represents the most prestigious marine scientists in the world, we'll get a good turnout. In addition to press releases, we're contacting everybody we know at major newspapers and TV stations."

Ted piped in, "Firefighters and the military should catch their attention, plus doubters can't claim we're a bunch of liberals. The press will get some great one liners."

Mark asked what I thought was missing. I smiled. An eminent climate researcher was asking my opinion.

"What about the difference between skeptics and doubters?"

Ted nodded. "What would you say?"

"Something like skepticism is part of the scientific process. All of us question our assumptions, data, interpretations. But global warming doubters are very different from skeptics. They already have their minds made up. They know the climate isn't warming, and they're out to prove it with recycled myths they don't question themselves."

Mark said, "Focusing on skepticism, that's terrific."

The others nodded.

I kept going. "I might want to challenge the press. Ask why they keep up with 'two sides' coverage when over two thousand climate change experts agree the climate is warming, it's mainly human-caused, and consequences are serious. Like the supersized storms we've had recently. Economically, how could we deal with a half dozen of those?" The words tumbled out. "Ah, and I'd ask why they continue to use so-called 'expert' skeptics whose work isn't even peer reviewed."

Over the next two hours we discussed details about the symposium.

When Tony and Mark were getting ready to leave, I felt comfortable enough to ask a favor.

"Before you take off, I need help from all three of you."

Tony said, "Since you agreed to speak at the last minute, how can we refuse?"

I described the NSF proposal with Gordy. "We need six scientists to work with the fishermen. If each of you were interested, I'd only have to find three more."

"Sounds intriguing and potentially important. I'm in," Mark said.

Tony agreed. "Me too."

Ted laughed and shook his head. "Like Tony said, how can I refuse?"

Ted walked his guests out to their cars and came back into the house to find me at the sink rinsing dishes.

"Hey, you don't need to do that."

"I'll wash, you dry."

It was quick work, and we were soon done. I pulled out an old ladder-back wooden chair, sat at the kitchen table, and ran my hand across the top, feeling cuts and grooves from decades of use.

"There's a lot of history in this house."

Ted sat down. "It's partly why I care so much about preserving the planet for people coming after us."

We sat, taking in the quiet of Ted's home.

I blurted out a question that came out of nowhere. "Has Harvey been here?"

"What?"

I had to come up with something fast. "Um. Harvey's into old houses. If she hasn't seen this one, she should."

Ted pulled his head back and stared at me. I decided it was time to leave.

22

My surroundings weren't familiar, and I pulled over. I'd been driving blindly, lost in thought. If Harvey and Ted were seeing each other and wanted to keep their relationship private, that was their business. There must be a good reason.

It was possible I was jealous that my best friend had a guy. I'd repeatedly told Angelo, Harvey, and Connor I was perfectly happy on my own. Maybe it wasn't true.

Usually I talked these things out with Harvey, but that felt too awkward now.

Perhaps my weird behavior had something to do with being nervous about the conference. In two days, I'd be a main speaker at a big-deal symposium. My body shook. I slid the seat back, bent over, put my head between my knees, and took deep breaths.

I sat up and whispered, "You can do this. You'll get through this." A quarter of an hour later, I could drive again.

I tried to distract myself, tried humming something happy. "Somewhere Over the Rainbow" popped into my head. I stopped in the middle of my third attempt at "Why, Oh Why Can't I?" There was something about this song. The memory came in a flash—my mother singing one phrase and me, ten years old, singing the next one.

The next morning found me driving to the University of Maine in Orono. On campus, I circled a green bordered by red brick buildings, parked, and walked to the biology building. Tony introduced me to my coach—an all-business thirty-something named Kristi, who'd teach me how to deal effectively with the press. Tony said he'd be back to take me out for lunch.

I followed Kristi down the corridor. A mix of hip and professional, she wore a black tailored blazer over skin-tight leather pants with matching high heels. In an empty classroom, we sat opposite each other in hard plastic chairs.

"I work with scientists who speak in jargon to other scientists," she said. "Since you understand one another, you don't realize how hard it is for most of us to fathom what the heck you're talking about."

She slapped a hand on the armrest. "That can lead to confusion, and worse, incorrect assumptions."

I raised an eyebrow.

Kristi went on. "Scientists speak in code. When you talk to the press, you must use plain language."

"I need examples from my field."

She stood, clicked up to the front of the room, and wrote "uncertainty" on the whiteboard. "What does this word mean to you?"

"It's a statistical term about probability and confidence in a value."

"And to non-scientists?"

I frowned. "Maybe 'I don't know?'"

"Good. How about the term 'random?'"

"Haphazard?"

"Or 'arbitrary' or 'accidental.'"

For the next two hours we worked on scientific terminology and came up with simpler ways to say and explain things. It was fun at first, but I was happy when Tony showed up. At lunch, it was a relief to talk jargon again.

The hardest part came in the afternoon. Tony and Kristi acted the part of the press and fired questions at me from their seats. At the front of the classroom, I struggled to field them.

"Dr. Tusconi, how can you say global warming's happening when we had such a cold winter?"

"Dr. Tusconi, there're weather stations near parking lots. Doesn't that mean the measurements are biased?"

"Dr. Tusconi, I've read the sun is heating up the earth. Is that true?"

It's hard to answer questions like these simply but with enough detail to address what's being asked. For instance, the idea that warming is caused by a hotter sun is a common misconception. NASA satellites show that solar activity has generally decreased as atmospheric temperature increased. Also, if the sun were getting brighter, we'd see higher temperature in both the lower and upper atmosphere. Instead, earth's surface and lower atmosphere have warmed, but the upper atmosphere has cooled. That's what you'd expect from the greenhouse effect, the result of a blanket of gases like carbon dioxide.

But it's damn difficult to briefly and simply explain all that.

I'd practiced not using jargon. So when asked about ocean acidification, instead of saying "pH is decreasing" I said, "the ocean is becoming more acidic."

Kristi had also urged me to use strong visual examples to make an idea clearer—like what gorgeous Pteropods look like when they dissolve in acidic waters.

I tried that out. "Pteropods are called sea angels because they fly through the ocean on what look like wings. They're lovely, delicate animals. It's terribly sad to watch these beauties turn into grotesque, misshapen creatures as their shells dissolve away in acid water."

"Not bad," Kristi said. "But instead of 'it's terribly sad,' say how it makes you feel."

That was going to be especially hard.

Kristi also urged me to talk about "real people" whenever I could. Like—"the seawater is already too acidic for oyster farmers on the west coast. Their oysters can't even grow shells."

On top of all of this, the questions came like water from a fire hose. If everyone talks at once, you can't scream "shut up so I can hear you!" You have to keep your cool, be polite, professional, calm—and throw in a bit of humor.

Uh-huh.

I drove back home utterly bushed. Hours answering questions made my jaw hurt. But Tony and Kristi were pleased. After a long trip, the drive down Water Street in Spruce Harbor always makes me smile. That evening the relief was palpable.

23

IOPENED MY KITCHEN DOOR and flipped on the light. My laptop was on the table, and out of habit I logged on to my email. Most messages could wait, but the subject line of one was "MOI Meeting," so I sat down and opened it. The message read:

Science Bitch. You're next.

Gripping the edge of the table, I read the words over and over—then slammed the computer shut. I shoved back my chair and walked into the living room, over to the picture window. The wind had picked up, and beach rose bushes on the lawn swayed wildly about. Beyond them all was pitch black.

The night matched my mood. The shock from the vile message on my screen had given way to anger. This was an outrageous, illegal invasion of my privacy. Whoever sent the message was sick, evil, or crazy. Maybe all three. The hairs on the back of my neck prickled.

I backed away from the window and slowly turned around. Until that moment, I'd never felt afraid in my home. I ran into the kitchen, pulled the cell phone out of my purse, and called Ted. While his phone rang, I bolted the kitchen door.

He answered on the fourth ring.

I glanced at the clock. It was 11:10. "Ted, I'm sorry if I woke you. It's Mara."

"No problem. What's the matter?"

"The climate change mafia. They're after me now." I recited the message. "I'm really pissed. How can these crooks get away with this crap? And how do you know if they're dangerous?"

"Well," he said, "these threats usually don't lead to anything. But just in case, are your doors locked?"

"I just threw the kitchen door bolt. The other door's always locked. What do I do with this email? Contact the IT folks?"

"Start there. But are you sure you're okay? Would you feel more comfortable if Angelo came over?"

"I'll be okay."

After passing the email message on to the IT people at MOI, I took a quick hot shower to relax and crawled into bed. The wind howled off and on all night.

I'd been at my desk for a couple of hours working on my talk when the phone rang. It was a technician from Maine Coastal Oceanographics. I'd sent them a temperature logger to be repaired when Seymour had denied my match. For the time being, I had to make do with equipment on hand.

"What sensitivity do you need?" the technician, Clay, asked. "What's the research?"

"We want better spatial and temporal coverage of water temperature in Georges Bank. It's global warming research."

For the next half-hour we discussed the data logger's specs. We finished, and Clay asked, "Hey, you're in Spruce Harbor. Know an engineer named Angelo De Luca?"

"He's my godfather."

"Great. Would ya say I asked about him? I knew Angelo at Shea Engineering when I started out. Left more'n ten years ago."

"Marine engineering's a small world."

"Yeah, his engineering group worked on gasket seals for scientific equipment, mainly subs."

My mouth went dry, and I stammered my response. "Submarine seals? Angelo worked on that?"

"Yup. He was head of structural engineering. Anyway, say hi."

I thanked Clay and dropped the phone into its cradle. This was not possible. My godfather couldn't have overseen structural engineering at Shea. The man responsible for my parents' deaths worked at Shea. He could very well have been in Angelo's department. If so, that meant that Angelo kept this from me—and that my godfather was partly to blame for what happened. He always said he raised me because he loved me. Maybe it was something else.

Work was out of the question now. I grabbed my purse, ran down the hallway and back downstairs, and wandered around the parking lot searching for my car. Zombie-like, I drove home. For the next two hours I prowled the house, sitting on the couch, the easy chair, opening the fridge but not taking anything out, going back to the couch.

Angelo was my rock. He was good. Honest. He wouldn't hold back such an important thing from me.

The phone in the kitchen rang. It was Ted.

"Mara, I looked for you at work."

"I didn't feel good and came home."

"You okay?"

"Yeah."

"Want to go over your talk?"

I had to get my mind off Angelo. "How about a walk on the beach? We can go through it there."

Ted pulled into my driveway. I met him outside, and we made our way down to the beach together. Ted updated me on Frank who now lived in his parents' house and visited the clinic as an outpatient. With everything else on my mind, I hadn't thought much about Frank.

"You know," Ted said, "the guy's really smart. I wonder if he's capable of hoodwinking his caretakers to get out of that institution. If so, he's a potential danger."

"Who'd be in danger?"

"He's got to harbor some strong feelings about you."

"Jeez," I said. "Put him on the list." I strode down the cobble beach.

Ted caught up. "Mara, what does that mean?"

"It means I don't want to worry about who's out to get me. I don't like feeling paranoid. I've always considered myself strong and independent. But with that email last night—"

"Are you sorry I got you involved? In giving the talk?"

We walked along in silence. I stopped, unsnapped my sandals and rolled up my capris, picked my way down to the water's edge, and waded in up to my knees. The cold water stung. I shuffled my way through the sand, back up to Ted.

"No, not sorry. I think about Stan. And Peter, of course. I'm not sure what I feel, to tell you the truth. Maybe stunned that researchers struggling to figure out what's going on with the climate are seen as bad guys—and that people with astounding amounts of money are behind that. Maybe discouraged so many people believe the doubter's lies that we're perpetrating a hoax. I feel small against the odds. Feeling small, that's lousy."

Ted put his hands on my shoulders. His touch was firm and his eyes indigo blue. "Mara, you're the last thing from small. You can make a real difference by speaking frankly and directly to the press. I don't think you realize the power of your voice and conviction. You're passionate in what you believe. Your integrity is absolute."

All I could manage was, "Okay then, thanks."

Ted stepped toward me. I stepped back.

"There's something else besides the threatening email. Guess it's what really set me off."

"What's that?"

I toed the sand. Angelo was a very private person, so I couldn't say who it was. "Someone I've trusted for a long time. I discovered something they should've told me. Something important. I thought they were truthful and now—"

We strolled down the beach. Ted said, "A while back I was engaged to someone I assumed was an angel. I found out she'd had an affair with one of my best friends. It was over, but she never told me about it. I was devastated and cut off the engagement. Later, I wondered if I'd done the right thing. People are complicated, Mara. They do things for reasons that seem right to them. I don't know. It's not as clear-cut as I thought." Ted kicked off his sandals. We made our way down to the shallow water, wading through it all the way back.

I had to talk to Angelo about Shea Engineering because it was eating away at me, so I invited myself for dinner. After Ted left, I tried to figure out how to broach the subject but still hadn't worked that out when my car came to a stop at the end of the gravel road on Seal Point.

I walked into the kitchen and could tell right away something was amiss. Angelo hadn't started dinner. Vegetables on the cutting board weren't diced and the table wasn't set. Worse was his appearance. His normally ruddy face was pasty and brown bags sagged beneath his eyes.

"What's wrong? What's happened?"

"Let's sit down," he said. No wide-grinned hello, no invitation for a glass of wine. This was going to be bad. I took the chair opposite him.

Angelo placed his hands on the kitchen table. He stared at them and said, "Do you remember my brother Marco?"

"Sure. I liked him. He made silly jokes."

"You remember he died of colon cancer."

The wake was a shock. Marco, who'd been so full of life, in his open casket. "Yes."

"Colon cancer's common in my family. It's good you and I aren't related." His attempt at a joke was strained. "I had a colonoscopy last week. I told you. Something they found was bad. They sent it out."

"When will you hear?"

"In a few days."

His eyes, stricken with anxiety, searched mine. I reached across the table and took his calloused hands in mine. The hands that'd helped me remodel my house and got greasy when he fixed my car.

"We don't know yet. It might be nothing. They're being cautious. They have to be like that."

He squeezed hard and let go. "You're right, dear. Let's keep thinking that." His voice was dull. "Did you want to talk to me about something?"

I couldn't ask him about Shea Engineering now and had to think fast. "Umm. Well, I've been worrying about that speech I have to give. I'm really nervous."

"It's funny you saying that. Your mother was an absolute wreck whenever she had to speak to the public."

"Really? I didn't know that. From what I saw she was, you know, charismatic. She really connected with people."

"Bridget Shea Tusconi was charismatic all right. She lit up a room, only partly because she was beautiful. On stage, she was poised, smart, totally in control. But nobody knew how she had to gin herself up. One time before a speech, she came out of the bathroom white as a ghost—she'd just emptied her stomach. Ten minutes later, she was behind the podium and nobody could've guessed."

I was in the audience when my mother spoke. She immersed people in a realm as foreign as the moon—moaning whales, crunching parrotfish, a fluid and three-dimensional world as dangerous for humans as it was fantastic.

And they loved her for it.

"That's astonishing."

"You know, I'm pretty sure I have something she wrote. Let's see if I can find it."

Angelo rummaged around in his office while I sat at the kitchen table, stunned. It was impossible to imagine that my famous mother was terrified about speaking in public. She'd given hundreds of talks about everything from the slaughter of whales to fish kills in the Chesapeake Bay. How could she have been afraid all that time?

Angelo came back, manila folder in hand. "I found it."

He looked over my shoulder as I ran my hand across the dusty surface, slipped my finger under the cover, and slowly opened it. It held two typed pages. Angelo slid them side by side. The piece was clearly a draft. In some paragraphs, words had been penciled in. In other places, my mother crossed out a whole phrase and wrote a new one in the margin in her neat longhand. I touched her penciled-in words.

"May I keep this?"

"Of course. Sorry I didn't think of it a long time ago. There's no date. She was writing an article for a women's magazine, I think. Something about facing up to your demons. She never finished it though. I don't know why."

I volunteered to get dinner going. Angelo took chicken and salad fixings out of the refrigerator. We reminisced about my mother, stories we'd talked about before. I kept glancing over at the folder on the kitchen table. It held my mother's words—her own voice describing a dread we shared.

That turned out to be a welcome distraction for Angelo and me. After dinner, he washed our dishes and I dried.

"Mom got nauseous. Maybe that's why I'm so prone to seasickness."

Angelo handed me a wet plate. "You inherited her beauty, spunk, smarts, and weak stomach. Guess you've got to take the bad with the good."

I patted his arm.

On my way out the door, I hugged Angelo extra hard. My last words were, "You're going to be okay. I know it."

What I felt, of course, was anxious as hell.

Back home, I saved my mother's article until my favorite time for reading—before I go to sleep. Sitting up in bed, I reached for the folder on my nightstand and began to read.

I've talked with many women afraid to speak in public. They're surprised I share this fear because I speak to audiences so often. This article won't give you tips such as what to wear. Instead, it's about being yourself.

We communicate with people every day. You might talk with your children about why baby birds die or your husband about his boss. In those situations, you're a skilled communicator. You explain your thoughts clearly and express complicated ideas. That's the person you want to be in front of a group.

We all want to connect with each other. And we want to listen to a speaker who is a real person—who's authentic.

That was followed by a mostly crossed out paragraph. Only a few words and phrases were legible—"look at your audience," "presence," and "credibility." I put the pages back on the nightstand. This wasn't about being prepared or memorizing your first sentence. It was more profound than that. I did have confidence in my ability to communicate with people in a wide range of situations. My students responded well to my teaching, and Harvey and I talked about personal problems. My mother was telling me to dig deep, find that communicator, and put her up there in front of the podium.

I reached for the pages again and read the final paragraph.

It's possible you're always going to be nervous—even sick to your stomach—before an important speech. I used to worry this was a sign of weakness. I don't think that anymore. I just deal with it the best I can and get on with what I believe is really important.

I flipped off the light, put the pages on the pillow next to me, and snuggled under the covers. My mother's phrases drifted through my mind, and I didn't feel so alone with my panic anymore.

If I'd been open with Angelo about my secret fear, he would've given me my mother's article years ago. Then her words could have helped me all this time.

Funny. I was hurt because I suspected Angelo had kept something important hidden from me. He probably hadn't done that at all. But I had—and paid the price.

24

THE NATIONAL ASSOCIATION OF OCEAN Scientists meeting was scheduled in MOI's largest auditorium in the early afternoon. Ted and I went through my talk over lunch in the Neap Tide.

"You guys went all out with publicity," I said. "Lines like 'scientist attacks climate change deniers.'"

He frowned. "That was Mark. I'm worried it's over the top." He fiddled with the saltshaker and tipped it over. "But we're dealing with desperate fanatics."

I righted the saltshaker. "I can't dwell on that now."

Ted made a tent with his fingers and, elbows on the table, leaned toward me. "Describe your talk."

When I finished, Ted said, "You don't go halfway, do you?"

"Full-speed-ahead kind of gal."

He shook his head and laughed. We both reached for the bill, but Ted got it first.

"Split it?"

"Least I can do. This isn't easy for you." Ted pulled out his billfold. I reached over and patted his hand. "You're a wonderful friend."

Ted opened his mouth and closed it again with the slightest shake of his head. With a quick "thanks," he got up and left, navigating the bustling room.

Back home the kitchen clock said 1:30. Over an hour to go. I changed into my professional-give-a-talk clothes—a classy black sleeveless dress with matching jacket. Twin silver barrettes kept my hair in place. To finish, I dabbed on lipstick.

On my way out, I stopped in front of the full-length mirror. Looking back was a slender woman with large green eyes, long auburn hair, and a full mouth. Very much the professional. I gave her a nod and left.

I arrived at MOI at 2:15. Walking around the parking lot, I tried deep breaths. But panic and nausea began to set in.

I mouthed my mother's words. *I just deal with it the best I can and get on with what's really important.*

With fifteen minutes to go, I headed over to the auditorium. Ted spotted me, and we walked along together.

"I know you're nervous. Think I'd be scared stiff. I've given dozens of scientific talks, but this is a whole other animal."

Ted had no idea.

He pulled open the auditorium door, and I stepped into a room buzzing with excitement. The size of the audience was flabbergasting. Probably about five hundred and more streaming in. As we walked to the front, my legs felt weak, and I touched a few seats for support as we went along. I met the other speakers, and we chatted until the session started.

Ted had invited Seymour to welcome the audience. To his credit, Seymour was a good speaker.

After thanking the organizers, Seymour said, "Thirty years ago, when I worked with marine biologists in Woods Hole, I first heard about global warming. But none of us thought the consequences—drowning islands, disappearing glaciers, and the rest—would happen in our lifetime."

As Seymour went on, I wondered which marine biologists he meant. I slid my smartphone out of my pocket and surreptitiously searched Seymour's name to check out who he published with back then.

One co-author on a couple of papers really puzzled me. I itched to ask Ted. Seymour had finished. I slipped the phone back in my pocket and applauded along with everyone else.

Mark was next. As he began, I turned around. Every seat was taken and latecomers had lined up along the wall. I swallowed hard and turned back. The big screen showed the iconic image of earth from space—the one that helped so many comprehend the fragility of our exquisite blue planet.

"Look at the vastness of the earth's oceans," Mark said. "Think about the enormous amount of heat required for the marine temperature increases we see right now. Despite what doubters say, there's *no* explanation for that scale of increase other than the human-caused greenhouse effect."

When he'd finished, Mark's applause was loud and heartfelt. I was pleased for him. Mark was a smart, hard-working scientist. So was Stan. A smart scientist in hiding because some powerful people wanted to muffle the research—and the researchers themselves.

I remembered Seymour's co-author and nearly gasped. At that moment—like the big win on a slot machine—it all clicked into place. That Peter's death was no accident, who was responsible—it all fit.

Now I desperately wanted to get Ted's attention, call Harvey—*something*. But in the front row of the auditorium, I was powerless to do any of that.

I was about to go on stage to give what might be the most important speech of my life. I had to put my realization aside and keep my focus on the task at hand. I closed my eyes, put other thoughts away, and listened to the speakers.

The firefighter was next. A tall, well-built guy, he was pretty spiffy in his forest service uniform. His first image was an eye opener—satellite pictures of California fires. I'd forgotten how shocking they were.

"Looks like all of southern California was ablaze, doesn't it?"

He ran video footage from the latest inferno in Colorado. The fire roared, and the scale of the burn and speed at which it spread was astounding. Firefighters in yellow jumpsuits and hardhats sprinted from flames heading directly toward them. You could almost smell smoke.

I glanced back. Everyone was glued to the screen.

He concluded with, "Climate change is here right now. Wildfire season starts much earlier, ends later. Fires are bigger, hotter, and they last a whole lot longer."

Only one more talk before mine. My mind drifted back to my new insight, but I forced myself not to think about it. I closed my eyes.

It's about being yourself.

My neck muscles felt like ropes.

We all communicate with people every day. That's the person you want to be.

The Army guy was next. With his colorful decorations and badges, he was even spiffier than the firefighter. He spoke right to the press.

"You guys've got to do your job to educate the American people about what's really going on. We in the military aren't a bunch of tree huggers. This thing is real, it's human caused, and it's a national security risk. Retired generals and admirals are working to alert the public and lawmakers that climate change is making the world a dangerous place. And you're in a position to help us."

Judging by the frenetic note taking going on, at least some of the press responded well to his call.

Oh my god, I was next. I had to walk all the way to the stage, up the stairs, and to the podium. I had to open my mouth and intelligent words had to come out.

We all want to connect with each other. And we want to listen to a speaker who is a real person—who's authentic.

As Mark introduced me, I stood. My legs held just fine. I yoga breathed, walked to the stairs, climbed the three steps, and strode to the podium. My heart pounded so hard people could probably see a mound going up and down on my chest. The image struck me as funny, and I grinned.

It's about being yourself.

An ocean of people looked at me expectantly. In the back, I spotted Connor and Angelo. I stood behind the podium and adjusted the microphone. I acknowledged the audience with a smile and began.

"It's wonderful that there are so many reporters here. If you don't mind, could each reporter raise their hand?"

A sea of hands emerged from the crowd.

I put my palm on my heart. "Wow. You give me hope because scientists who study climate change greatly need your help."

A familiar cartoon appeared on the screen. "I'll read this for people in the back. The reporter on the left says, 'Five thousand eleven out of five thousand twelve scientists believe global warming is human caused.' His boss on the right answers, 'Get the one guy for your interview.'" I waited for the laughs to subside then said, "I'd like to dissect one word in this cartoon—'believe.' As a scientist, I don't *believe* or not believe in what I study. Instead, I try my best to understand the evidence. It's what the data tell me. You could say, I suppose, that I believe in data.

"Data change over time. Today, we have buoys that report water temperature every minute at stations along our coast. Those numbers give us more confidence that warming in coastal Maine waters is a repeatable pattern.

"But still, confidence is not a *belief.* That's why we use terms that drive reporters crazy—words like *probably* or *most likely.*

"That's completely different from climate change doubters. No matter what, they *believe*—in fact, they *know*—they're right.

"And unlike scientists, some of these believing doubters are in the driver's seat."

An image appeared on the screen.

"This Midwest senator heads up one of the most powerful environmental committees in the country. Bear with me a minute while I read the exchange on the next slide. Here, a *Déjà View* magazine reporter questions the senator at a U.N.

meeting in Australia.

"*Senator, there are lots of really smart people here. They're experts in policy, government, science, you name it. Seriously, they've all been duped?*

Absolutely, he said, nodding his head vigorously.

So this climate change thing is a well-organized scam?

Like I said, yes.

Who is behind this global scam?

Those tree-hugging radicals are one powerful force, sonny. Even got movie stars behind 'em.

What movie stars?

Joan Rivers, he responded."

A few people in the auditorium tittered but most stared at the exchange open-mouthed.

"The senator walked away before the reporter could tell him that Rivers was dead."

That got a laugh.

"Seriously, I urge you reporters to discuss the implications of this exchange on your own. I just ask—and I'm going out on a limb here—that you consider your children and grand-children's future."

I explained why reporters seeking different perspectives on climate change should choose scientists who work for organizations conducting primary research on climate science and who published this work in peer-reviewed scientific journals.

"To end, I'd like to dedicate this speech to my friend and colleague Stan Huntley, a famous climate change scientist from MIT. A few days ago the FBI urged Stan to go into hiding because a well-financed climate change doubter group hacked Stan's email and threatened to kidnap his kids."

I turned toward the smiling image of Stan on the screen.

"Stan, wherever you are, I pray your nightmare ends soon."

I waited for the auditorium lights to come up again. The room was silent—and I went cold with the idea that I'd blown it. Suddenly, the audience exploded. Some even stood up. Way in the back, Connor clapped so hard I thought his hands might fall off.

Mark thanked the speakers and audience and announced the press conference a half-hour later. It was fifteen minutes before I could leave the auditorium. People crowded around us. Many who spoke to me were scientists who wanted to say how much they appreciated my honest and blunt talk.

I found a bathroom and splashed water on my face. My speech had been a delicate balance of accuracy and persuasion—not at all what I was used to—and it took an awful lot out of me.

At the press meeting, the speakers sat at a table in front of a smaller room and fielded questions. Most were expected, and I had a chance to talk about skeptics and believers. Afterward, people from the press stayed in the room so they could meet on their own.

For me, it was finally over.

Ted, Mark, Tony, and I found each other in the corridor. Their smiles and thanks were all I needed to know I'd done a good job.

Ted said, "We're heading out for drinks and an early dinner. Want to come?"

"Sorry, I'm absolutely exhausted. But I'll walk you out."

On the way, I whispered to Ted. "There's something I have to tell you."

Ted's eyes followed the others, now walking away from us. "Can't it wait, Mara?"

I suppressed a flash of irritation. Ted just wanted to enjoy a well-earned achievement. "Sure."

He sped off to catch up with the other two.

25

As I drove the winding road to my house, all I could think about was a hot shower, glass of wine, simple dinner, and bed. I also had to call Harvey to tell her about my revelation.

I stepped out of the car and slammed the door.

The blow to the back of my head hardly registered.

I came to on my back with my hands tied behind me and feet bound together. My head pounded with every heartbeat. Panic swelled up from my gut. I forced it back down and focused on what I could understand.

Blinking, I looked around. I was on the backseat of what looked like a van. It was still light and the corner of my garage was just visible. The van was in my driveway.

Closing my eyes, I drifted off.

A door slid open. I woke with a start and squirmed up to a seated position. A bulky guy with a cruel grin peered in at me.

"Hey. Look who j-joined the livin'."

It was Jake—Liver Eyes—from *Intrepid*. His T-shirt strained against his bulk, and a mass of wormy tattoos spread down his arms and up his thick neck. Someone stood behind him.

Jake climbed into the van and leaned his bulk over the seat in front of me.

"G-g-got my orders, lady. I'm g-gonna tell you what'll happen. So ya s-stew on it."

I stared up at him.

"Out in the ocean we're g-gonna dump ya."

He leaned closer. His breath stank of beer and cigarettes.

"D-drownin'. Know what it's like? Ya try to hol' your b-breath, but ya c-can't. Ya s-s-uck in w-water. Lungs burn. Ya c-claw to the top. Th-then it all g-goes dark."

The van spun.

I was going to drown in the ocean. Just like my parents. I shut my eyes. The spinning stopped.

The other person spoke. A man. "She's got the message. Leave her be for now."

The Irish accent was unmistakable. It was Ryan, *Intrepid*'s first mate. My friend.

Jake backed out and slammed the door.

Ryan. What was Ryan doing with a thug like Jake? I pictured the Irish seaman. His tweed cap, blue eyes. On *Intrepid*, we chatted about the country across the sea he loved—his altar boy days, the Irish farm where he grew up, Ireland's economic ruin. He taught me knots. I helped him understand my research.

It was impossible to imagine why Ryan would want to hurt me.

Jake said that someone gave these guys "orders." Maybe my insight a few hours earlier was right. And maybe they were behind everything—Peter's death, the latest "accidents" on the ship, the hate email.

I could hear the garage door open and pushed myself up a bit. Jake and Ryan came out carrying my shorter kayak— the red one Ted used only days earlier. I fell back down on the seat. Jake yanked the van's backdoor open and slid the boat across the seats and against the far window. It just fit.

Why they wanted my kayak was too much to think about. Exhausted by nausea and the driving headache, I drifted off again and woke when doors opened, slammed shut, and the vehicle came to life.

Bumping along in the back of the moving van, my emotions were an untidy mess. One moment chilling terror poured through me. Next, I was so angry I wanted to crawl over the seat and—what? Bewilderment settled in like drying mud. How long the trip took, I don't know. The motion of the van lulled me in and out of consciousness. Awake, my head hurt horribly and mouth was sandpaper dry.

Horrific images flashed by. My parents desperately trying to get out of the sub as seawater poured in. I felt their fear. Maybe they held each other as the world went dark. Did they think of me then?

Anguish washed over me. I was such a goddamn fool.

My anger with Angelo was petty. Now cancer might kill him without me there.

Then there was my secret. For so long I'd told myself, and nobody else, that I couldn't speak to the public about the ocean, its mysteries and threats. Maybe the Tusconi reputation paralyzed me. But if I'd trusted Angelo and confessed my fear, he'd have given me a treasure—my mother's writing.

All along, I had the ability to follow my parents' example, to make a bigger difference in peoples' hearts and minds. Now it was too late to prove I could do it again.

Do something. Stop feeling sorry for yourself. Focus on what you might have control over.

The van turned off a highway, slowed, and came to a stop. A door opened and slammed shut. I sat up. Ryan and I were alone.

"Ryan," I croaked.

He turned his head and looked back—his eyes big and haunted.

"Dr. Tusconi, I didn't know—" he whispered.

I cut him off and whispered back. "Later. Help me?"

He scanned the woods in front of the van and nodded. "Jake?"

"Taking a crap. Couple of minutes. And he's got a gun."

"Phone?"

"I'll look for Jake's."

I prayed.

"Got it," he whispered.

"Call nine-one-one. Police. Say where we're going. *Now.*"

He bent down and spoke softly to what I guessed was a dispatcher.

Jake returned to the van well after Ryan finished the call. Everything would be okay. The police would save me.

The van rolled down the road for a while, turned, and stopped.

Doors opened, slammed shut. It was still light, and I sat up. Out the window: water, sand, some trees. No police yet. Ryan and Jake were gone.

26

RYAN SLID THE DOOR OPEN and made his way back to me. He undid my ankles and hands and handed me a wetsuit. I squinted at the thing. It was *my* wetsuit.

"Put this on. Quick. I'll sit on the running board."

"Police?"

"Don't know. Just do it."

Pulling a dress off and a wetsuit on in the back of a van isn't easy. Since I was so stiff, it'd be even harder.

I looked around. No police, no sirens.

Cool it. They'll be here.

I kicked off my shoes and pulled my dress over my head. Muscles screamed. I grabbed the wetsuit and glanced at Ryan. I needed answers.

"Ryan, why?"

Ryan sounded close to tears. "Dr. Tusconi. Swear to God, I didn't know it was you."

"But why?"

It came out fast. "I made an awful, awful mistake. It's my mum. She needs money real bad. The farm in Dingle. Hundreds of acres, ours for generations. The economy's tanked."

I shoved my legs into the wetsuit. "They paid you money to kill Peter, and you did it?"

"No! The winch. That never should've happened. You've gotta believe me. It was supposed to scare you all. Not kill anyone. Lord, not kill anyone."

"But Jake's said I'll drown. You're helping him."

"I didn't know. They'd tell MOI if I didn't help with a

little job. That's what they called it. When we got to your house, I 'bout died."

I pulled up the wetsuit. "Who's behind this?"

"Shh. They're comin' back."

"Ryan. I'll do what I can for you. Where are the goddamn police?"

"Don't know. Get out."

I zipped the wetsuit, shuffled up, and stepped out. Pebbles cut into my bare feet.

Two men strode toward us. Jake was in front, the other right behind him.

The second man stepped forward. "Hello, Mara."

I was right. It was John Hamilton.

But this was not the man I knew. John Hamilton was a neat milk-toast. This man had greasy strings of hair spilling down his forehead to eyes black like tar.

He reached forward and jiggled the tab on my wetsuit zipper. Revulsion, acid sour, bubbled up from my gut. I stepped back.

Pull yourself together. Buy time.

"What're you doing, John?"

"Getting rid of you."

"Because I found out about Frank?"

He spat out words, rapid fire. "You're not gonna ruin my sweet deal."

"Did Pacific Petroleum make you rich?"

Hamilton narrowed his eyes. I must have guessed right.

"What about the perfect local fuel?"

"Finally wised up. Algae research took forever. Frank had it, then didn't."

I had to keep Hamilton talking. "You hired Jake and Ryan to scare scientists on *Intrepid*?"

His grin was diabolical. "Nobody was supposed to die." He glanced at Ryan. "But it turned out okay."

Hamilton could've been talking about a turn in the weather.

"And you targeted MOI because the lobster and hot acidic water research got lots of press?"

"Real good money to stop it."

I was about to ask another question when he cut me off. "Enough. Jake, tell her the deal."

Hamilton pulled a gun out of his pocket that he tossed from hand to hand as Jake talked.

"A mile out we t-toss ya over, then your b-boat where ya c-can't get it. You'll be d-dead 'n t-two minutes."

It was a smart plan. Someone would find my body and boat. The authorities would figure I drowned in cold water. Smug, John Hamilton looked at me like I should applaud.

I gave him the best blank face I could muster.

"Jake," he said, "get her kayak."

Hamilton gestured toward the beach with his gun. "You first, *Dr.* Tusconi."

A twenty-five-foot powerboat floated at the water's edge. The tide was low and the engine on the stern tipped up out of the water. I picked my way over pebbles, feigned tripping, and took a quick look back. Nobody. My stomach clenched. The police should already be here.

Behind me Hamilton barked, "Hurry the fuck up."

Jake passed me, my kayak under his arm, and marched toward the boat.

I stepped over dried seaweed at the high tide line. Flotsam or jetsam? Angelo had quizzed me about the difference. Angelo. He had no idea where I was.

At the waterline, Hamilton barked more orders. "Jake, slide the kayak into the boat. Then stay here. Ryan, get in.

"You," he shoved me forward, "in the boat."

I stepped into the water. Glacial. I pictured Hamilton's sneer as he threw me over. Arctic seawater would stream into the wetsuit and feel like icy nails boring into my head.

Stop it. I clambered into the boat.

Hamilton pointed to the bow with his gun. "Sit right there."

He climbed aboard and made his way back to the stern. Ryan stood at the console and scanned the shore.

Jake shoved us off. When we reached deeper water, Ryan used the hydraulics to lower the motor.

Hamilton yelled, "Start the goddamn thing."

Ryan turned the key. Nothing. He tried again. Still nothing. He turned to Hamilton.

"Got to check on somethin'." Ryan walked back to the motor, leaned over, and looked underneath.

Hamilton snarled, "You get this thing going or—"

Ryan straightened up and stared at Hamilton's gun. "I'm doin' my best here."

He was pale, sweating. Scared.

We'd drifted out, and I looked back to shore. There was the road and van. No police cars. Just scowling Jake.

Ryan backed away from Hamilton. When he reached the console, he turned and tried the key again. The motor sputtered and caught. Ryan slid the throttle. I held on as he steered the boat into deeper water. We sped forward, away from hope.

Seal Harbor, down the coast from Spruce Harbor, was too shallow for most boats and quiet even in summer. In early spring, it was deserted. Perfect for Hamilton's purpose. The lobster boat that approached us from the north was a surprise.

From the stern Hamilton yelled at Ryan, "Why's that out here?"

Ryan shrugged and yelled back. "Maybe he's landin' lobster south of us somewhere."

"Stay away from it."

Ryan turned toward open ocean. The lobster boat changed its course in the same direction. I squinted and tried to make out who she was. In the early evening light, it was impossible to tell.

As we slapped through waves, I bounced hard on the bow and stole peeks at the retreating shore. Jake was still alone.

When the distance between us and the approaching boat was less than a quarter mile, I squinted again. The lobster boat had shifted course, and I could see her starboard side.

No. It couldn't possibly be.

There was no mistaking the graceful sweep—Cape Island style, dark blue.

She came closer. New railings. Rusty scuppers.

Gordy's *Bulldog*.

27

Hamilton snarled, "Steer clear of that boat." Ryan turned. *Bulldog* stayed her course, which gave me a view inside the cabin. In his orange overalls and ratty baseball cap, Gordy Maloy stood at his wheel, alone in the open cabin.

Hamilton yelled, "Keep this course." He patted his pocket. "No funny business."

Hamilton, Ryan, and I fixed on *Bulldog*. It looked like the two boats were going to pass each other.

Gordy swung *Bulldog* hard around, kicked her into high gear, and headed right for us. Fast.

"That guy's crazy!" Hamilton hollered.

I flattened myself on the deck.

Ryan froze.

Hamilton held onto the stern railing. "Do something!"

Ryan wrapped both hands tight around the steering wheel. With Gordy still headed right for us, Ryan swung the boat to starboard, to port, starboard again, trying to dodge *Bulldog*.

It didn't work. Only a hundred feet away now, Gordy mirrored Ryan's arcs and bore his boat down on us.

In the back of the boat, Hamilton held onto the railing, leaned over, and shrieked at Gordy. He fired a couple of wild shots with his gun.

I was sure *Bulldog* would ram our boat and couldn't figure out which side to jump off.

Suddenly, Gordy slammed his boat into reverse.

I rolled over. Ryan craned his neck back toward the stern. Hamilton reached out over the port railing like he was going to grab Gordy and strangle him.

Ryan held onto the console and rammed his throttle forward. Then he grabbed the wheel and wrenched it to port with a fierce yank.

Hamilton let out a high-pitched scream—a scared-shitless scream—as he sailed over the railing and hit the water. The sea drowned his shrieks as he went under. Thrashing like a spiked shark, he came up, squealed, and went down again.

I leapt to my feet.

This man couldn't just die. He had to go to prison. I was the one in a wetsuit—I steeled myself for the icy water.

Gordy yelled, "Stay. I'll take care of the friggin' bastard."

Ryan maneuvered within twenty feet of *Bulldog*.

Hamilton slapped at the water, like he was trying to push himself up. Either he couldn't swim or his wet clothes weighed him down. He gasped, "Get me out! The cold! The cold!"

His motor chugging, Gordy pulled closer and threw a life jacket onto the water. Hamilton grabbed the preserver, held it tight to his chest, and managed to kick his way over to *Bulldog*.

He looked up at Gordy and sobbed, "Help me. Please."

I called out, "Watch it, Gordy. He's got a gun."

Gordy leaned over and glared at Hamilton.

The panicked man craned his neck up and howled, "No! No! It fell out. Swear to God."

Gordy reached for the arm of his hydraulic lobster pot hauler and rotated it over the side of his boat. He called down, "Got a belt on?"

Hamilton swallowed some water, coughed, and nodded.

Gordy looped a line through the pulleys, tied a hefty hook at the end, and lowered it. "Hook your belt."

Hamilton tried to grab the hauler, arms flailing and splashing as he took on more water.

Gordy growled, "Release the damn jacket and hook your belt or swear t'God I'll leave you out here."

Hamilton grabbed the line, pulled it down, and secured it.

Gordy switched on the motor. Hamilton squeezed his eyes shut and clung to the line as he slowly rose out of the water. When the man was halfway up, Gordy reached down and wrapped his hands around Hamilton's wrists. Finally, the dripping scoundrel was level with the rail, and Gordy flipped the switch on the motor. Hamilton's head rolled back like he was losing consciousness. Gordy reached into Hamilton's pocket, pulled out the gun, and shoved it into his shirt pocket. He rotated the hydraulic arm back into the boat, lifted Hamilton over the railing, and undid the hook. Like a sodden rag doll, Hamilton fell into the boat, out of view.

Ryan shouted, "Gordy, I'll ride back with you and watch the bastard. Mara can handle the wheel."

I took Ryan's place and maneuvered the powerboat parallel with *Bulldog*. With practiced ease, Ryan stepped onto our stern rail and into Gordy's boat.

I called out, "You good?"

Some of the old Ryan came through in his crooked grin. "I'm real good. See you back there."

I pulled away and pushed the throttle forward. The powerboat leapt toward shore. Soon, Jake was visible at the water's edge. Hands in the air, he backed away from something higher up on the beach in the shadow of the trees. Another person came into view. I slowed the boat to a crawl so I could take it in.

Harvey held her Weatherby rifle to her shoulder, the sight squarely on Jake.

"Looks like you got a real prize," I yelled.

Police cars screeched to a stop by the van. Officers jumped out. Harvey dropped the butt of the rifle, engaged the safety, and positioned the gun across her chest. A male officer ran over and talked to Harvey, then signaled a female officer who stood beside Jake. I almost applauded when she cuffed him.

Behind me, a lobster boat chugged.

Gordy called out, "Mara, ferry us to shore."

I sidled up to *Bulldog* as the sunset to our west colored the sky pink. Ryan and Gordy lifted the mumbling Hamilton over the lobster boat's railing, dropped him into the powerboat, and climbed in. I beached my boat. A couple of cops ran over to Gordy, who acted like they were long lost friends. As Gordy started to fill them in, I jumped onto the sand. A policeman strode up to me.

"Mara Tusconi?"

"Yes, sir."

"Glad to see you. You okay?"

"Officer, I'm *thrilled* to see you. I'm okay."

"We've been on a goose chase." He gestured toward the powerboat. "If it hadn't been for Gordy there, we'd still be at Eel Harbor."

"*Eel?*"

"Yeah. Where the dispatcher directed us."

Oh my god. Because of his stutter, Jake sometimes dropped the letter S at the beginning of a word. As a result, *Seal* Harbor must have become *Eel* Harbor when Jake told Ryan the van's destination.

"But how did Gordy—?"

"Let's talk later. There's an ambulance waiting for you behind our cruisers. I'll walk you up and take a statement at the hospital."

For the second time in a week, I traveled via ambulance to the Spruce Harbor Hospital. Ryan and I rode together. John Hamilton, strapped in a gurney, was in another ambulance. I didn't know where Jake ended up and didn't care.

Above the siren, the paramedic asked, "Can I cut the wetsuit? I've got to check your blood pressure."

On my back I said, "Cut away."

After he'd finished, the medic let me sit up. Ryan sat on the other bench.

"Ryan, I know nothing about sentences for unintentional manslaughter in Maine. But I *will* do what I can for you. You helped me escape and John Hamilton get captured. That's got to count for something."

With the excitement of the rescue behind him, Ryan had returned to a dismal state. Head in his hands, his voice quaked as he spoke. "They'll lock me up. No sky, no sea. And I deserve it."

There was nothing to say. I reached over and put my hand on his arm until we reached the emergency entrance.

The officer found me in my cubby room. I was tucked under three blankets with an IV tube dripping into my arm. He took a chair beside the bed while a female officer stood next to the door.

"How're you doing?"

I turned toward him and winced. "In good shape, they say, considering. The ER doc was worried about the blow on the back of my head, but the CT scan looked okay." I glanced at the needle in my arm. "I'm dehydrated but beginning to warm up."

"I'd say you're damn lucky."

"You bet. Ryan—"

"Hold up, let's start from the beginning."

Soon, he had my signed statement with details about the kidnapping, up to his arrival at the beach. "Tomorrow, assuming you're okay, come down to the station. I'll have more questions." He stood to leave.

"Before you go, how did Gordy help you find us?"

"Like I said, we raced over to Eel Harbor, but nobody was there. Gordy barreled in with that lobster boat of his. He yelled that you had to be at Seal, not Eel, Harbor and that he was going to steam over there. So we jumped back into the cruisers and got to Seal as fast as we could. It's a good ten miles between the two harbors, you know, and the roads aren't great."

"But how did he know it was Seal?"

"Something about the number of moorings. You'll have to ask him."

"One last question. Where's Ryan now?"

"The ER folks checked him out and said. he could go. He's on his way to the station."

"If it weren't for Ryan, I wouldn't be here."

He held up his hand. "You've made that clear. He'll be arraigned tomorrow in front of a judge. They'll decide what's going to happen to him—in the short term anyway."

A nurse opened the door and stuck her head in. "Your godfather is here. Ready to see him?"

My grin answered that question.

Angelo appeared in the doorway. He strode over to the chair beside my bed, sat and took my hand, and rested his forehead on it. When he looked up, his eyes were damp. "My God, Mara, you gave us a scare."

I squeezed his hand. "I'm so sorry. What about your biopsy? I've been so worried."

The tired smile said it all.

"Negative."

It was my turn to tear up. "I'm so very glad."

I asked, "How 'bout Harvey?"

"She left the waiting room to get something out of her car, but she'll be back. Connor, Gordy, and Ted are still out there."

"Ted?"

"Yes indeed. That young man appears to care a great deal about you."

I didn't have the energy to explain that Ted and Harvey were a couple. "You know, he probably feels responsible for what happened."

"Uh-huh. Now, tell me what happened."

I filled Angelo in. He didn't need to hear a few things— like the blow that knocked me out.

My description of Harvey made Angelo chuckle. "You should've seen her. Rifle to her cheek, she looked like Annie Oakley. And Jake so scared he'd fall back into the water to get away from her."

Angelo patted my hand. "Enough. You must be exhausted."

"What time is it?"

"Little after ten."

Hours since we left the house. It felt like days.

I closed my eyes for a moment. Sleep would be bliss. "You're right. But maybe I could give a little wave to the trio in the waiting room?"

The nurse returned with Connor, Gordy, and Ted. "These gentlemen say they want to see you smile, then they'll leave."

I did more than smile. I giggled at the sight of them standing side by side at the far end of the room. Gordy held his dog-eared Boston Red Sox cap across his heart. With tight curls and red cheeks, Connor looked like a grownup altar boy. And as usual, Ted's wayward lock of hair fell across one eye.

"Tomorrow I'd love to spend time with each of you. But now—"

The nurse ushered the three men out of the room. She returned with another visitor. "Absolute last one."

Harvey walked over to the bed and shook her head as she looked down at me. "Damn, girl, we've got a lot to catch up on."

"Sure do. One question before you go. How'd you end up on the beach with Jake in your scope?"

"Police scanner in my truck. I heard 'Eel Harbor,' 'kidnapped female,' took a good guess, and was on my way when 'Seal, not Eel' came through. I was minutes away and beat the police. That asshole Jake was down by the water. He threatened me, so I grabbed the rifle. The bully turned to mush in a second."

28

I BLINKED MY EYES OPEN and squinted at trees outside the window. Ash. Odd. No ash trees grew near my house.

A familiar voice cut through the confusion.

"Hello, dear."

I turned my head.

Angelo stood beside the hospital bed. A rumpled shirt stuck out of his pants and heavy stubble said a.m. not p.m.

"You sleep here?"

"No way I could at home. How do you feel?"

I shifted my position. "Okay. My head hurts."

"Doctor said you were hit from behind."

So much for keeping anything from Angelo.

The next day I visited the doctor, who said I was doing well, and spoke with a few reporters about my talk. In between, I managed to plead my case for Ryan, who was released on bail. Ryan gave the money John Hamilton paid him to the Catholic Charities. To help Ryan's mother, Gordy and Connor said they would raise an equivalent amount from some Irish organizations.

I even managed to submit the short version of our grant proposal to the National Science Foundation on time. If the program officer liked it, Gordy and I could submit the full proposal. I checked the prose for the tenth time, clicked "submit" on NSF's website, and sent a prayer and the grant into cyberspace.

I was flabbergasted when the program officer got right back to me.

At the end of the day, Harvey dropped by my house with some maple syrup.

"This spring's run right from a sugar shack."

"Thanks, Harv, but it's not like I'm sick."

"Let me spoil you a bit. Walk on the beach?"

Out on the sand, Harvey said, "I called Sarah. This afternoon is good for her. We can meet there."

"How did she sound?"

"Tired—but anxious to learn what you discovered about Peter's email."

"I'm not sure how much detail to go into."

"Keep it simple. She's got a lot on her plate."

"I've had time to think about what's happened since Peter's death. It's changed me, at least some."

I stopped, gazing into the unusually glassy water. "I'm going to try not to jump to conclusions. It's embarrassing to think back. I took Ted for Frank, was convinced Georgina Hamilton was the crook instead of John, and thought Angelo was deceitful. I didn't tell you about that."

"You can fill me in later."

"The second thing might be even more important." I turned toward her. "The people I trust most—I need to let them help me with my hardest problems."

"Like—"

"You didn't even know about this one. Fear of speaking to the public. That paralyzed me. I put it aside and tried to forget about it."

"You should give yourself a break there, Mara. Living up to your parents' legacy was a huge burden. Hardly anyone has to deal with that."

"We've all got secret burdens, Harv."

Back in my driveway, Harvey said she'd see me later at Sarah Riley's house.

I joined Gordy in the Lea Side, Spruce Harbor's townie bar. We had some excellent news to celebrate.

Gordy asked our waitress for his third beer while I nursed my champagne. "Tell me again what she said, the, um, what'ya call it lady."

"National Science Foundation officer. Our preliminary proposal was creative, right on target, the best in the lot. She wants a full proposal in a month. Assuming all goes well, we'll get the award in late summer."

The waitress slid Gordy's beer across the table. He took a slug. "Damn. Who'd think? Doc and me—a team."

I tapped my glass against his mug. "Gordy, you saved my butt. We're more than a team. We're cousins. We're buddies."

Gordy grinned. "An Irish friend is a friend forever."

On my way to Sarah Riley's house, I went through the maze of events that gave us insight about Peter's email. Painting a simple picture would be hard.

I parked behind Harvey's truck and walked up the shelled pathway. When Sarah did that, Peter would never be inside to greet her.

I followed Sarah into the kitchen as she said, "The twins are with their grandmother so we can talk."

I pulled out a chair and wiggled out of my jacket.

Harvey said, "I've told Sarah a little, but why don't you start from the beginning."

Sarah folded her hands on the table and nodded. She looked drained, and I hoped our discovery would give her a little relief.

"It took me a while to contact the sci-fraud organizer. Lots of phone calls. Finally, I reached a woman who Peter talked to about, um, his discovery."

Sarah leaned forward. "She talked with him?"

"Yes. In his email, Peter said he was suspicious about a genetic engineering claim, which we thought might be medical or agricultural. Instead, the woman said it had something to do with sustainability."

Sarah frowned. "Like what?"

"It was local—Peter was going to visit, you'll remember. We had a lucky break with a newspaper piece about an aquaculture company up the coast. They claimed they created a super-seaweed."

"And that was it?"

I nodded. "Peter didn't know, but he discovered an elaborate hoax. One of the most powerful oil corporations in the world tried to undermine local energy production."

Sarah's hand flew up to her mouth. "Is that why Peter died?"

"It's related. It looks like these folks were going after MOI climate change researchers in general."

"But this oil corporation. Won't they continue, you know, to harass you all?"

"Great question, but we don't think so. It's a high profile company. Their local agents did very stupid things. We're guessing they got carried away and went off on their own. The oil conglomerate will distance themselves from the whole debacle as fast as they can."

After our visit, Harvey and I leaned against her truck.

I looked toward the house. "Do you think that helped?"

"The idea that Peter died fighting the big battle? That's given war wives solace for centuries."

At the end of the day, I enjoyed a juicy development that involved Seymour. The MOI director, Frederick Dixon, invited me to his office. I was surprised Seymour was there as well.

I shook the director's hand. "Dr. Dixon. It's been a while."

"Yes, yes, Mara."

Dixon gestured to chairs around his cherry table. "Please sit down."

I did. Seymour picked the farthest chair from me.

Dixon peered at me. "Mara, I'm so proud of the wonderful work you do educating the public. The climate change research here at MOI is cutting edge, as you well know. Not many researchers have your dual talent—great oceanographic science and excellent communication skills. I knew your parents. They'd be proud of you too."

Dixon beamed. Taken aback, I cleared my throat. "Well sir, I appreciate your saying that. As you can imagine, it's a struggle to balance the two."

"It certainly must be."

Turning to Seymour, Dixon said, "And I want to thank you for your support of Mara. In the spirit of the Tusconi professorship you hold, I'm sure you've helped her."

Seymour looked like he'd swallowed a fly. He recovered quickly. "Ah, Biological Oceanography appreciates your recognition of our efforts." He didn't look at me.

It was the perfect moment to practice something Harvey urged me to do—be political. I said, "Seymour, I appreciate your help."

Seymour's mouth opened and closed like a guppy's, but nothing came out. Dixon walked me to the door. He winked as he shook my hand again. MOI's director was a pretty shrewd guy.

29

ON FRIDAY AFTERNOON, GORDY ORGANIZED a "spring cruise and picnic." When I arrived with chicken salad sandwiches, Harvey, Connor, Ted, and Angelo were already down at the town dock.

The day was unusually warm with light winds. I wore a spiffy new dress for the occasion.

Canvas bags filled with food, drinks, clothes, and blankets cluttered the dock.

"You'd think we were going away for days," I said. "*Bulldog*'s not that big."

Angelo pointed seaward. "Here comes Gordy, and that's not *Bulldog*."

Gordy stood at the wheel of the fifty-foot yacht heading our way.

As the boat sidled up to the dock, Connor called out, "What's this?"

Gordy climbed down from the bridge. "Step aboard, ladies and gentlemen. And Connor—I've got friends."

Harvey took Gordy's hand as she stepped off the dock. "Where are we heading, captain?"

Gordy yanked down his baseball cap a bit. "A little island an hour's steam from here. Great place to picnic."

With everyone and everything finally aboard, Gordy steered away from the dock. The yacht slid through the water, and I joined Harvey at the stern. We were past the harbor's headlands before either of us spoke.

"It's been, what, two weeks since we stood on *Intrepid*'s deck as she pulled away from Spruce Harbor?" I said. "So much has happened."

"Let's see. Peter died, Frank nearly ran you over and then attacked you in a parking lot, Jake tried to sabotage the second cruise, Stan went into hiding, you gave the speech of your life and then got kidnapped—"

I took over. "Ryan called the cops who got lost, John Hamilton nearly drowned, the cops arrested Hamilton and Jake plus Ryan, I ended up in the hospital, Angelo doesn't have cancer, and it looks like Gordy and I will get that NSF grant."

"And you think *I'm* one tough babe, Tusconi?"

I patted her hand. "No way I could've managed without you."

Connor walked up and stood next to Harvey. "I heard part of that. Maybe you two should stay out of trouble for a little while."

Harvey laughed. "With Mara, that's not so easy. Hey, it's getting cool out here." She slipped a jade-colored shawl off her shoulders and draped it around mine. "Great dress, Mara. Green brings out the color of your eyes. You'll need this though."

The shawl was deliciously soft and warm. Cashmere, of course.

I stroked Harvey's gift. "It's lovely, but what about you?"

"I've got another like it in my bag." And with that, Harvey took Connor's hand and together they headed for the sheltered cabin.

As Connor held the cabin door for Harvey, a grin spread across my face. Well, well. Harvey and *Connor* were a couple. And why not? Connor wasn't much older, they were terrific people, and they both loved to hunt. What more did you need?

We reached the island, and Gordy anchored offshore and shuttled us to shore in a dinghy. The dinghy only held three, and Harvey and I went first. The boat bottomed out in the island's shallows. We pulled off our shoes, hiked our dresses up, jumped out into cold water, and yelped. On the beach we spread the blanket and waited for the others.

Harvey brushed sand off her legs. "What you said yesterday about secrets…well, there's something I want to tell you."

"You okay?"

"I'm fine. It's about my past and involves Ted."

"Now I'm really curious."

"I'll tell you the end first." She leaned back on her elbows. "Ted's my half-brother."

I stared at her. "What?"

"Hold on. It's complicated. I only found out a few years ago. Growing up, I knew Ted as my Midwest cousin. I hardly saw him—the occasional wedding, that type of thing. A few months before Mother passed from cancer, she told me the family secret."

"Your cousin was your half-brother?"

"Yes. He was grown by then. My father had an affair with his brother's wife. Ted was the result. The couples reconciled and decided Ted should be raised with my uncle's family."

"Damn. When did Ted learn about this?"

"When he turned twenty-one."

"It's a coincidence you're both oceanographers."

"I had a hand in that. Ted was in high school when I got my PhD. At a reunion, he asked if a guy from the Midwest could go to college and be an oceanographer. He got A's in math and physics and was bonkers about the ocean. I encouraged him to follow his dream."

I glanced over at the yacht. Connor and Angelo were in the dinghy.

"Why didn't you tell me before?"

"Ted wanted to apply for the MOI job—it was perfect for him—which put me in an awkward position. Of course, I told Seymour the situation. Since I hardly knew Ted, he decided it wouldn't be a conflict of interest. But I couldn't be involved in the hire in any way."

"Why not tell me after he got the job?"

"I wanted you to get to know him on his own merits, not as someone related to me."

"I probably would've thought about him differently. Hey, let's help unload."

As we strolled toward the approaching dinghy I said, "I pictured you and Ted together. Now I know why. You look alike for a reason."

We walked the beach, ate too much, and walked the beach again before Gordy announced it was time to head back.

"Wait," Connor said. "I haven't heard Mara explain how she figured out John Hamilton."

I turned to Gordy. "Is there time for this?"

"Go ahead."

The group formed a semicircle around me. Harvey and Angelo already knew what I was about to say. They listened politely anyway. "I should've figured it out sooner. But Hamilton played his role perfectly. He was the hen-pecked, meek guy devoted to sustainable biomass production. He did make one mistake. During my Sunnyside visit, I asked him details about algae—basics, like the growing medium they used. He claimed ignorance, and I figured he was the business end of the operation. In his welcome speech at the meeting, Seymour mentioned colleagues from Woods Hole days. I was curious, so before he finished I searched the net and found papers Seymour co-authored back then. John Hamilton was first author on several. Each focused on growing algae in large volume systems."

Connor asked, "Which meant he knew answers to your questions?"

"Absolutely. Once I saw through his deception, everything else fell into place. Like the money. Turn John Hamilton into a scoundrel and the answer jumps out at you. What better way to undermine local sustainability fuel research than create a high-profile, bogus operation—secretly funded by a wildly wealthy, unscrupulous petroleum company—that eventually becomes a huge embarrassment for the whole sustainability industry?"

Connor frowned. "Local fuels threaten oil companies?"

"The algal biomass industry claims they could produce enough petroleum for the U.S. on a fifth of the land used to grow corn."

"Still," Angelo said, "why did he say he didn't know about growing algae?"

"That was clever. When the fraud was made public, he could claim ignorance about the whole thing. Put the blame on Frank."

Gordy looked toward the yacht. "Folks, we've got to go."

"Couple more questions," Connor said. "Why kidnap you? And why target MOI?"

"Hamilton blames Frank's psychosis on me. Besides that, he intended to announce the fraud—Frank's fraud—on his time schedule. Months from now, maybe longer. Then I showed up with the isotope data. Hamilton had no idea how many of us knew the truth about his super-seaweed but realized the hoax would be exposed soon. So his big money would disappear much earlier than he'd planned. He hated me for all that. And me dying of hypothermia in my wetsuit—he believed he'd come up with a perfect murder untraceable to him."

"And MOI?"

"That's harder to understand, but the guy is nuts and obsessed. Climate change deniers harass well-known climate scientists. MOI's a leader in the field, especially in the northeast. So we were a handy target right down the coast from Sunnyside. Unfortunately, his brand of harassment got out of hand."

"Of course, scientists in other controversial fields have been personally attacked," Harvey added. "Like ones who work with animals."

To that Connor raised an eyebrow and gave Harvey a worried look.

I rode with Connor in the dinghy and asked, "What will happen to Hamilton and Jake?"

"They must be in Maine's state prison. Kidnapping, attempted murder? They'll be locked up for a long time."

"And Georgina Hamilton?"

"From what you said she's smart. Assuming she's innocent, she'll divorce her husband and get away as fast as she can."

In the stern, Gordy operated a little two-stoke. He snorted. "She's probably long gone."

Connor straddled his middle seat so he could talk to Gordy. "Fast thinkin' there. Figuring out it was Seal Harbor."

"Christ. 'Course it was Seal. Eel's full of boats and nosey boaters. Seal's like a graveyard this time of year."

It was almost my graveyard.

On the way back to Spruce Harbor, I pulled Harvey's shawl tight around my shoulders and settled down on the bow of the yacht to watch the sunset. I wasn't there a minute when Ted walked up. "Mind if I join you?"

I patted the storage box I used as a seat. "Plenty of room. We're out of the wind here and the view's terrific."

We bounced through the waves in silence for a while. Reverence for quiet was something I loved about Ted's company.

We passed a few islands not far from Spruce Harbor. I said, "Harvey just told me your family's secret."

"Ah."

"When you found out who your dad really was—I can't imagine."

"A real shocker, to say the least. But it was good they waited. I was old enough to understand such things happened."

"I guessed you and Harvey were close, you know, but I had the nature of your relationship wrong."

"Oh?"

What did "oh" mean? I glanced at Ted for a hint but got none. The yacht rounded an island and headed due west, directly toward sunset. Streaks of purple, orange, and vermilion splashed across the sky.

The sight stunned me so much that Ted's hand on mine didn't register until we'd crossed through Spruce Harbor's headlands. A couple of weeks earlier, I would have slid my hand away.

Now, I only leaned back and let Gordy take us home.

Acknowledgements

I WOULD NOT HAVE BEEN able to write this novel without the help of many talented and generous people. The SeaScape group heads the list. Connie Berry, Judy Copek, and Lynn Denley-Bussard were with me every step of the long journey. I thank the wonderful folks at Torrey House Press, especially Kirsten Johanna Allen and Anne Terashima, for professional help, faith in me, and commitment to conservation literature. I have benefited from the considerable experience of my agent, Dawn Dowdle. Linda Vescio, Noy Holland, Kathy Whelan, Sandra Neily, and Ramona Long offered helpful suggestions on early drafts. Sisters in Crime and especially Guppies (the Great Unpublished) Agent Quest group helped me navigate the domain of queries, pitches, agents, and publishers. I appreciate financial help from Mystery Writers of America's McCloy award for new writers. CrimeWave, Maine Publishers and Writers Alliance's annual meeting, has been a terrific way to meet Maine mystery writers. I am grateful to climate fiction (cli-fi) enthusiasts, especially Mary Woodward who runs our cli-fi site. Dan Bloom, who coined "cli-fi," encouraged me to keep plugging along.

Finally, John Briggs, my husband, offered constant help with great ideas and much needed humor.

Scientists and others will note that I do not use metric measures. I believe this makes the story more accessible. Scenes aboard the oceanographic research vessel are necessarily limited in technical detail. I have never heard of a buoy accident like the one featured here but describe other marine phenomena, including ones related to ocean warming, as accurately as possible for readers.

Charlene D'Avanzo
December 2015

About Charlene D'Avanzo

Marine ecologist and award-winning environmental educator Charlene D'Avanzo studied the New England coast for forty years. Today she uses mysteries to immerse readers in Maine waters' stunning beauty and grave threats. An avid sea kayaker, D'Avanzo lives in Yarmouth, Maine. *Cold Blood, Hot Sea* is her first novel.

SHADOW SPIRIT
OF
THE SEA
A MARA TUSCONI MYSTERY

CHARLENE D'AVANZO

Forthcoming July 2017 from Torrey House Press

TORREY HOUSE PRESS

SALT LAKE CITY • TORREY

1

A T LIGHTNING SPEED, THE UNFEELING sea swept me from safety and any living soul.

Toward the North Pacific Ocean. Vast. Frigid. Deep.

Fear bubbled up from my belly and coated my tongue—taste of cold metal. I slammed a neoprene-gloved hand against the jammed lever at my hip. Did it again. Glanced back.

The offending rudder sat limp and useless on top of the kayak's stern.

"Worthless piece of *crap*."

Without a working rudder on a long, skinny kayak, I was helpless against the relentless current and wind. I'd slip right past Augustine Island, our campsite for the night. I'd be alone on a darkening seascape in a cockpit inches from deadly cold Pacific water three thousand feet deep. The Aleutian Islands, to the west, were a thousand miles straight shot.

My fragile vessel was on the edge of nowhere.

Vigilant about tipping over my skinny boat, I used the paddle as an outrigger, twisted a bit, and squinted at five kayakers paddling away from me. Bouncing on waves and fading fast from view, they looked tiny against the blue-black sea. Harvey Allison was easy to spot in her banana-yellow boat. That's where I should be—paddling alongside my fellow oceanographer and best friend. We'd talk over salmon counting methods the Haida used on rivers in the archipelago, a Canadian national park.

Best laid plans. When the Haida watchman who'd motor us out to Augustine was stuck on another island with a broken outboard, we joined a group of kayakers for the trip from Kinuk Island to Augustine.

Taking kayaks was my idea. "It's only a two-mile paddle. Piece of cake. We'll have plenty of time to sample fish in Eagle River."

I didn't factor a runaway kayak fifty miles off the coast of British Columbia.

My gazed shifted from Harvey to the guide at the front of the kayak pod. The metal taste in my mouth turned hot. A half hour into our expedition, this Bart character left me—in *his* piece of crap defective kayak? Harvey must have pleaded with him to get me. But he'd tell her he couldn't leave four novice kayakers scared and paddling poorly. Maybe he figured I was an experienced kayaker who wanted to go off on her own.

If he did, it was Davy Jones' locker for me.

I swiveled back gingerly, careful not to flip the boat. In forty-five-degree water, hypothermia would paralyze me in minutes. I'd be dead in fifteen. Squaring my body against the foot braces, I dipped the paddle into the water. Right, *pull.* Left, *pull.*

Hit Augustine Island. Just hit the blessed island.

Right, *pull.* Left, *pull.*

With every stroke, I willed the boat to swing toward that precious bit of land. But at high tide, the current carried me at racing speed—in the wrong direction. Worse, the kayak banged up and down against building waves, making rudderless steering a joke.

Fear bubbled up again like vomit. I spat it out.

Get a grip, Tusconi. Figure out how to fix the broken rudder.

I tried to picture the mechanism. Stainless steel cable ran from the rudder to foot braces inside the cockpit. The lift line cable at my hip allowed paddlers to raise and lower the rudder. So what could go wrong?

Lots. Lift lines jammed. Cables broke. Rudder brackets twisted. Things I could repair on dry land. But out here in the cockpit there was absolutely nothing to do but try like hell to reach that island.

Right, *pull.* Left, *pull.*

Back at the Maine Oceanographic Institute, I'd looked up currents and sea-state in Kinuk Bay. Both were treacherous oceanside of Augustine Island. Longer than wide, Augustine Island protected Kinuk Bay from the ravages of the Pacific Ocean. Once a paddler reached the outer bay, the current would race them toward the island's tip. If I slid past Augustine, the jig would be up. Crashing waves twice the length of my boat, icy water, sharks—I'd be at the mercy of all of it.

I swallowed hard, shut my eyes for an instant in an attempt to stay calm, then popped them open only to see that a kayaker's nightmare had suddenly materialized ahead.

Fog.

Makes sense. Water and air warmed by July's sun meets frigid open-ocean.

In moments, I was smothered in clammy fog—understanding why didn't help a bit. My heart pounded against layers of fleece, a paddling jacket, and life vest. Breathing quickly and hard, I swiveled and squinted in every direction. The vapor was so dense, there was no telling where murk ended and sea began. Waves booming against rocks said the island was tantalizingly close, but fog made me dead blind.

The kayak raced on through the gloom. Too soon, the booms were muted. The kayak had slipped past my refuge. That's when the certainty—I *would* drift out onto the vast ocean in a seventeen-by-two-foot boat—truly sank in.

My own gasps echoed back to me in the fog. Panicked, I was hyperventilating.

Calm down. Think._

I shoved the useless paddle into bungee straps on the deck, closed my eyes, and tried to breathe more slowly.

Thoughts of home drifted through my mind. If I died out here, what would people think?

That I drowned in the element I devoted my life to and loved. I could see the headline in my Maine hometown's *Spruce Harbor Gazette*—"Local scientist dies on sea kayak trip in Queen Charlotte Islands, Haida Gwaii, off coast of British Columbia."

A bizarre notion wormed its way into my thoughts. So bizarre I stopped paddling. Did this happen for a reason? Had the Raven—the native Haida's trickster—set me up to drown?

"*No*," I said aloud. "Spiteful spirits don't exist."

If someone wanted me dead, the trickster had to be human.

The fog thinned a bit. I looked back to see boulders at the tip of Augustine Island at least a half-mile to my stern.

I turned to face my ocean fate, raised my paddle overhead, and shrieked at the Haida spirits, Jesus, any damn thing that might hear me.

2

THE CALL, HIGH AND PIERCING, answered from behind. I
twisted toward it and nearly fell out of my boat. Close
on my port side, something long, sleek, and blood-red flew
by. Behind it an oval eye, black as liquid night, stared and
morphed into a glowing sphere. The rest was a black blur
longer than my kayak. Stranger still, the thing radiated a
warm glow that evaporated an envelope of fog around it.

In an instant, the thing was gone. I blinked. Only fog
to port. I craned my neck to starboard and behind as far as
I dared. Nothing.

*Fog and fear is messing with my mind. Creatures three
times my height don't exist except in fairy tales.*

Using my paddle as an outrigger again, I struggled
against waves that smacked the boat, fear racing through
my veins. A distant sound wormed its way into my consciousness. I jerked to attention and strained to hear it over
sloshes and smacks.

The friendly drone of a motor.

Fog amplified sound—especially low frequency, long
wavelength tones like foghorns.

And boat motors.

I closed my eyes, slowed my breathing, and focused
every synapse on the possibility of a distant low rumble.
Nothing but the slap of my damn boat against the water.

Heard it again.

An unmistakable drone grew louder. Impatient, I tried
to turn toward it and came close to tipping the boat. The
beat-up dinghy pushed through the fog and slid alongside,
motor sputtering. A man, hair blue-black, eyes intense like

a raven's—but very much human—sat in the stern, one hand on the outboard's handle. With the other, he swung a paint-chipped oar over his gunwale. I grabbed my end of the oar and our boats touched.

From peril to human contact in a flash—that was fast, even for me.

I gasped then managed, "Goddamn."

Silent, my rescuer studied my face.

"Who *are* you?"

His voice, deep and deliberate, carried a Canadian lilt. "William. Kinuk Island watchman. I was scheduled to take you over to Augustine but got stuck in Rose Harbor when the darn motor wouldn't start. I'm *so* sorry."

"Hey, marine motors take a beating. Thank goodness you're here now." I eyed his boat. Grayish bilge water sloshed over William's cracked rubber boots and generations of peeling paint colored the dinghy's interior green, white, and blue. Of course, my tub was no prize either. I pointed my thumb over my shoulder. "Rudder's stuck."

William pushed my boat forward. The kayak wobbled as he fiddled with the rudder. From his mumbles, I guessed the thing was good and jammed.

Finally, he yanked the offending appendage down into the water and let go. I positioned my left foot on the brace, swept my paddle across the water's surface, and grinned as the kayak swung to the right and faced him.

"By the way, I'm Mara. You saved my life. I'm in your debt—"

William held up his hand and, as if he rescued floating maidens every day, acknowledged my gratitude with a slight nod. "Mara, head back to the bay. I'll follow until you're out of the fog and safely on your way to the Augustine campsite."

In front of the puttering dinghy, I paddled hard and plowed through three-footers with ease. When we reached the calm waters of Kinuk Bay, William called out, "You okay now?"

I grinned and hand-pumped my paddle overhead.

"Great. See you on Augustine. I'm helping Bart with the kayak group." William cranked his motor. The dingy wallowed and he sped off with a roar.

I followed his wake. Alive, in control, and feeling strong, I drank in the seascape as the kayak slid through blue-green ripples tipped in silver that reached the western horizon. To port, a flank of moss-draped cedars hid the rest of Augustine's dripping forest.

As my kayak slid up the rocky shingle beside the other boats, Harvey ran over and pulled it higher onto dry land. She sat on the bow to steady the kayak while I climbed out.

Harvey skidded down the shingle and put her hands on my shoulders. Her gray eyes searched mine. "Mara. You vanished. I was so worried. What—?"

I gestured toward William's dinghy. He'd dragged it above the waterline. "William. Saved my butt."

Harvey looked at the banged-up dinghy and back at me, her eyebrows like question marks. "William? You mean that attractive young man who just arrived?"

"That's him, and he's twenty years younger than you."

Even though she looked my age—thirty-two—Harvey had just celebrated her fortieth birthday.

She tossed her blond, expertly shaped bob. "Thanks for the reminder. I'll remember it when you blow out eight more candles."

"Seriously, Harv. Sorry to worry you." I walked to the stern of my kayak. "My rudder got stuck." On both knees, I peered at the rudder and moved it up and down. "Don't see what's wrong with this thing." I opened my mouth to tell Harvey about the mirage but closed it. The time wasn't right.

"Your rudder didn't work. So, *what* happened?"

The aroma of grilled fish wafted our way. Down the beach, a couple of women stood around a campfire eating sandwiches. My stomach reminded me it was past lunchtime.

"Remember that current we talked about at the interface between Kinuk Bay and the open ocean? Twenty-plus knots? I got caught in it. The tide raced out and took my rudderless kayak with it."

"Good Lord! What—"

"Tell you more later. Okay? We need to talk about what we're doing, and I'd really like to eat something warm."

We picked our way to the campfire across loose rocks that covered the shingle beach.

"Same plan?" I asked.

"Yes. William's done the salmon counts, so he can explain the Haida's methods. We'll motor over to the river, review their procedures, and look at data. I've already talked to him."

We reached the campfire. Three women introduced themselves.

One of them, Gwen, said, "Sounds like you folks have work to do. Bart's taking us for a hike. We'll see you later."

William handed me a plate. "Have a salmon burger and take some chips. Are you alright now, Mara?"

His dark eyes locked with mine for an instant.

I took the plate. "Um, terrific. Thanks to you."

"Good. When you're finished, we'll take the boat around to Eagle River. There's a hut where we store our equipment. You can use any of it, whatever you want."

William held the boat while we stepped in. Given the grungy bilge water, I was happy to be wearing paddling wet shoes. From the way Harvey tiptoed to the stern, I was sure she thought the same thing.

As we motored along Augustine's eastern shore, William told us about our destination. "As you know, Eagle flows into the Pacific on the west side of the island. We call it the river that never sleeps because salmon run pretty much year 'round."

I yelled above the drone of the motor. "What's running now?"

"Coho. That's what we're counting."

We rounded the point where I'd screamed like a crazy woman and saw the bizarre vision an hour earlier. Both seemed like scenes from a movie. I shook my head to erase the image.

William beached the boat at the mouth of Eagle River. We hopped out, looked upstream, at the Pacific, back upstream.

"I've never seen a trout or salmon stream like this," I said.

William tipped his head to the side. "What do you mean?"

"In Maine, most rivers and streams that carry salmon are rocky and run fast. This looks like a wide, slow-moving, meandering river with a pebbly bottom."

"Eagle flows fast after a rainstorm, then it slows down."

"Makes sense. Augustine's watershed's small."

"William," Harvey said, "Why don't you give us a quick overview of the iron, ah, experiment."

William's eyes lit up. "Sure. We chartered a ship and added the iron out there." He gestured toward the Pacific. "Last week was the third time."

"Iron slurry, right? How much?" I asked.

"Yes. Two hundred tons."

I'd read the number. But standing on the beach, it was hard to imagine anyone dumping that volume of slurry into waters off the stunning archipelago. "That's a lot."

"Mr. Grant said we needed that much."

Roger Grant, in my opinion, was a slimy businessman who'd conned the Haida into giving him a lot of money for a bogus scheme. I'd have to figure out a polite way to voice my concern.

"We'll talk more about this later. But give us a quick overview of what Grant claims about iron and your salmon," Harvey said.

William tipped up his chin. "You know. Iron fertilizes the ocean, more salmon run up the rivers, and we make more money selling them."

Harvey nodded. I looked at the ground. We'd challenge the idea with the Haida environmental council the following day.

"I understand the U.N. hired you, Dr. Tusconi, and Dr. Allison to study the iron fertilization. Why is that?"

Noting the "Dr. Tusconi," I slipped into my professional oceanographer's voice.

"Plus Dr. Ted McKnight. He's flying into Kinuk today with our research equipment for the cruise. We're all from the Maine Oceanographic Institute. The U.N. considers dumping tons of iron sulfate into the ocean an international violation. They want an unbiased team, people not from this region, to visit the site and also duplicate some of your sampling to verify the results. We can talk more about this tomorrow on the ship when Ted's there."

William's fists were clenched. "But—"

"Lots of work to do now, William."

He relaxed his hands. "Alright." William pointed to a shed above the high tide line. "I'll go over how we count fish."

We followed him up the beach. "We want to get in the water and survey a reach ourselves," Harvey said. "When was your last count?"

"Yesterday."

"Good. We'll compare what you got with our data from today."

"The number of fish changes a lot from day to day."

Harvey and I glanced at each other.

In the shed, William showed us equipment he and an assistant used on the river. He picked up a device with wheels and revolving metal cups. "We have two current meters."

I nodded. "Stream gage?"

"It's upstream. Each time we count fish, we record river height."

In the corner, a wooden box overflowed with dive gear—wetsuits, booties, hoods, gloves, masks, and snorkels. Clipboards, waterproof paper, and markers lay on a shelf above.

"And your counting methods?" Harvey asked.

"We enter the river from downstream and slowly move up. Whoever's in the water calls out data to the other person walking alongside. Numbers of salmon and length. We estimate size from a ruler attached to a diver glove."

"One person covers the full width?" I asked

"Yes. The river isn't that wide."

"Time of day?"

"After sunset and on moonless or cloudy nights."

Harvey looked at her dive watch. "It's about four now. Let's go back, eat early, and be here by seven. Sunset's around eight."

TORREY HOUSE PRESS

*The economy is a wholly owned subsidiary of the
environment, not the other way around.*
—Senator Gaylord Nelson, founder of Earth Day

Torrey House Press is an independent nonprofit publisher
promoting environmental conservation through literature.
We believe that culture is changed through conversation
and that lively, contemporary literature is the cutting edge
of social change. We strive to identify exceptional writers,
nurture their work, and engage the widest possible audience;
to publish diverse voices with transformative stories that
illuminate important facets of our ever-changing planet; to
develop literary resources for the conservation movement,
educating and entertaining readers, inspiring action.

Visit **www.torreyhouse.com** for reading group discussion
guides, author interviews, and more.